ALSO BY DERRICK JENSEN

Railroads and Clearcuts

Listening to the Land

A Language Older Than Words

Standup Tragedy (live CD)

The Culture of Make Believe

The Other Side of Darkness (live CD)

ALSO BY GEORGE DRAFFAN

Railroads and Clearcuts

The Elite Consensus

The Global Timber Titans

STRANGELY LIKE WAR

THE GLOBAL ASSAULT ON FORESTS

DERRICK JENSEN

GEORGE DRAFFAN

Foreword by
Vandana Shiva

A POLITICS OF THE LIVING BOOK

CHELSEA GREEN PUBLISHING
WHITE RIVER JUNCTION, VERMONT

Book design by Peter Holm, Sterling Hill Productions
Printed in Canada

First printing, August 2003
10 9 8 7 6 5 4 3 2 1

Printed on Legacy Trade Book Natural, supplied by Webcom, a 100% post-consumer waste recycled paper.

Library of Congress Cataloging-in-Publication Data
Jensen, Derrick, 1960–
Strangely like war : the global assault on forests / by Derrick Jensen and George Draffan.
p. cm. — (Politics of the living)
Includes bibliographical references (p.) and index.
ISBN 1-931498-45-8 (alk. paper)
1. Deforestation—History. 2. Deforestation—Environmental aspects. 3. Clearcutting—Environmental aspects. 4. Forest ecology. I. Draffan, George, 1954– II. Title. III. Series.
SD418.J46 2003
333.75'137–dc21
2003055279

Chelsea Green Publishing Company
Post Office Box 428
White River Junction, VT 05001
(800) 639-4099
www.chelseagreen.com

CONTENTS

FOREWORD

The forest has always been my teacher in peace, in diversity, in democracy. Diverse life forms, small and large, moving and immobile, above ground and below, with wings, feet or leaves, find their place in the forest. The forest teaches us that in diversity lie the conditions of peace, the realization of democracy.

I grew up in the lap of Himalayan forests. I shifted from a research career in physics to environmental research and activism when the peasant women of my region started the chipko (hug the tree) movement.

Forests have always been central to India. They have been worshipped as Aranyani, the goddess of the forests and the primary source of life and fertility. The forest as a community has been viewed as a model social evolution. The diversity, harmony and self-sustaining nature of the forest formed the organizational principles guiding Indian civilization; the aranya samskriti (roughly translated as 'the culture of the forest' or 'forest culture') was not a condition of primitiveness, but one of conscious choice. According to Rabindranath Tagore, the distinctiveness of Indian culture consists of its having defined life in the forest as the highest form of cultural evolution. In *Tapovan,* he writes:

> Contemporary western civilization is built of brick and wood. It is rooted in the city. But Indian civilization has been distinctive in locating its source of regeneration, material and intellectual, in the forest, not the city. India's best ideas have come when man was in communion with trees

> and rivers and lakes, away from the crowds. The peace of the forest has helped the intellectual evolution of man. The culture of the forest has fueled the culture of Indian society. The culture that has arisen from the forest has been influenced by the diverse processes of renewal of life which are always at play in the forest, varying from species to species, from season to season, in sight and sound and smell. The unifying principle of life in diversity, of democratic pluralism, thus became the principle of Indian civilization.
>
> Not being caged in brick, wood and iron, Indian thinkers were surrounded by and linked to the life of the forest. The living forest was for them their shelter, their source of food. The intimate relationship between human life and living nature became the source of knowledge. Nature was not dead and inert in this knowledge system. The experience of life in the forest made it adequately clear that living nature was the source of light and air, of food and water.

As a source of life nature was venerated as sacred, and human evolution was measured in terms of human's capacity to merge with her rhythms and patterns intellectually, emotionally, and spiritually. The forest thus nurtured an ecological culture in the most fundamental sense of harmony with nature. Such knowledge that came from participation in the life of the forest was the substance not just of the Aranyakas or forest texts, but also of the everyday beliefs of tribal and peasant society. The forest as the highest expression of the earth's fertility and productivity is symbolized in yet another form as the Earth Mother, as Vana Durga or the Tree Goddess. In Bengal, she is associated with the sheora tree *(Trophis aspera)*, and with the sal *(Shorea robusta)* and asvathha *(Ficus religiosa)*. In Comilla she is Bamani, in Assam she is Rupeswari. In folk and tribal cultures, trees and forests are worshipped as Vana Devatas, forest deities.

But the forests, our sacred mothers, our teachers of peace and security, are themselves becoming the victims of war. It is a war unleashed by the violence of the monoculture mind, which reduces nature to raw material, life to a commodity, diversity to a threat, and views destruction as "progress." In *Strangely like War,* Derrick Jensen and George Draffan open our eyes to the terrorist assault on our living guardians and the destruction of our real security.

Vandana Shiva
August 8, 2003

NOTE

The pronoun *we* should be used only by royalty or those with particularly active intestinal flora. In this book, *I* refers to the primary author, Derrick Jensen, and *we* refers to both authors.

Deforestation

> It was strangely like war. They attacked the forest as if it were an enemy to be pushed back from the beachheads, driven into the hills, broken into patches, and wiped out. Many operators thought they were not only making lumber but liberating the land from the trees.[1]
>
> *Murray Morgan, 1955*

The very day we wrote the final words of this book, scientists declared that yet another subspecies of tiger had gone extinct in the wild (with only captives remaining, so discouraged they're dosed with Viagra to try to make them breed). Gone extinct. Such a passive way to put it, as though we know no cause, can assign no responsibility. It's almost as though we were to say that victims of murder passed away, or that victims of arson decided to move.

The South China tiger joins its cousins the Caspian tiger, Bali tiger, and Javan tiger, all victims of logging, roadbuilding, and the leveling of forests under this excuse or that.[2] The other tigers will almost undoubtedly join them soon.

It doesn't matter much to the tigers whether the forests are cut because Mao decided that "Man must conquer nature," or because the World Bank decided that "Man must develop natural resources." The forests are cut, the tigers are dead.

The forests of the world are in bad shape. About three-quarters of the world's original forests have been cut, most of that in the past century. Much of what remains is in three nations: Russia, Canada, and Brazil. Ninety-five percent of the original forests of the United States are gone.

We don't know how fast the surviving forests are disappearing. We don't know how many acres are cut each year in the United States, nor how much of that is old growth. We have estimates, and we'll give them throughout the book, but the paucity of information even on present levels of cutting reveals more than it hides: it reveals how desperately out of control is the whole situation.

The United States Forest Service and the Bureau of Land Management sell trees from public forests—meaning they belong to you—to big timber corporations at prices that often do not even cover the administrative costs of preparing the sales, much less reflect full market value. For example, in the Tongass National Forest in southeastern Alaska, 400-year-old hemlock, spruce, and cedar are sold to huge timber corporations for less than the price of a cheeseburger, and taxpayers pay for the building of the logging roads as well. The Forest Service loses hundreds of millions of dollars a year on its timber-sale programs. In other words, if you pay taxes, you pay to deforest your own land.

If you live in the West, Southwest, South, Northeast, Midwest, Alaska, or anywhere else in the United States where there are or were forests, chances are good you've seen or walked clearcuts, sometimes square mile after square mile, cut, scraped, compacted, and herbicided. You've seen lone trees silhouetted on ridgelines, and you've seen once-dense forests reduced to a handful of trees per acre. You've suspected and later learned that these few trees were left so the Forest Service and big timber corporations could maintain that they did not clearcut this particular piece of ground. And maybe you came back another time and saw that the survivors, too, were gone.

You've probably driven highways lined by trees, then pulled over to look around, only to discover that just like in old westerns, where false fronts hid the absence of real stores, you've been sold a bill of goods: a few yards of trees separate the road from yet more clearcuts. This fringe of trees, which reveals recognition on the part of timber corporations and government agencies that industrial forestry requires public deception, is common enough to have been given a name: the beauty strip.

Do yourself—and the forests—a favor. Next time you fly over a once-forested region on a clear day, look down. Pay attention to the crazy quilt of clearcuts you see below, to the roads linking clearcuts and fragmenting forests, roads that wash out in heavy rains to scour streambeds and destroy fisheries.

Only 5 percent of native forest still stands in the continental United States. Four hundred forty thousand miles of logging roads run through National Forests alone.[3] (The Forest Service claims there are "only" 383,000 miles, but the Forest Service routinely lies, keeping double books—a private set showing actual clearcuts, and a public set showing some of the same acres as old growth—misleading the public by labeling clearcuts "temporary meadows," reducing the stated costs of logging roads by amortizing them over a thousand years, and so on).[4] That's more road than the Interstate Highway System, enough road to drive from Washington, D.C. to San Francisco one hundred and fifty times. Only God and the trees themselves know how many miles of roads fragment the forests.

The forests of this continent have not always been a patchwork of dwindling and increasingly isolated natural communities. Prior to the arrival of our culture, unbroken forests ran along the entire eastern seaboard, leading to the cliché that a squirrel could have leapt tree to tree from the Atlantic to the Mississippi, never having touched the ground. Today, of course, it could still never touch ground, but instead walk on pavement. Polar bears wandered as far south as the Delaware Bay; martens were "innumerable" in New England; wood bison cruised that region; passenger pigeons passed overhead in flocks that darkened the skies for days at a time, Eskimo curlews did the same; rivers and seas were so full of fish they could be caught by lowering a basket into the water. American chestnuts ran from Maine to Florida so thick on the dry ridgetops of the central Appalachians that when their crowns filled with creamy-white flowers the mountains appeared to be covered with snow. Before European "settlement"—read conquest—of America, there was no such thing as "old growth," no such thing as "native forest," no such thing as "ancient forest," because *all* of the forests were mixed old growth, they were all native, they were all diverse, ancient communities. Difficult as all of this may be to imagine, living as we do in this time of extraordinary ecological impoverishment, all of these images of fecundity are from near-contemporary accounts easy enough to find, if only we bother to look.

Worldwide, forests are similarly under attack. One estimate says that two and a half acres of forest are cut every second. That's equivalent to two football fields. One hundred and fifty acres cut per minute. That's 214,000 acres per day, an area larger than New York City. Seventy-eight million acres (121,875 square miles) deforested each year, an area larger than Poland.

The reasons for international deforestation are, as we'll explore in this short book, similar to those for domestic deforestation. Indeed, those doing the deforesting are often the same huge corporations, acting under the same economic imperatives with the same political powers.

Apologists for deforestation routinely argue that because preconquest Indians sometimes "managed" forests by setting small fires to improve habitat for deer and other creatures, industrial "management" of forests—deforestation—is acceptable as well. But the argument is as false and unsatisfying as the beauty strips, and really serves the same purpose: to divert our attention from deforestation. This is analogous to saying that because someone once clipped a partner's fingernails, it's okay for us to cut those fingers off.

I saw this argument presented again just today in the *San*

Francisco Chronicle, in an op-ed piece by William Wade Keye, past chair of the Northern California Society of American Foresters. He wrote, "Native peoples managed the North American landscape, cutting trees and using fire to perpetuate desirable forest conditions. There is no reason that we cannot equal or better this record of stewardship."[6] Actually, there are many reasons. Indians lived in place, and considered themselves a part of the land; they did not come in as an occupying force and develop an extractive economy. They did not participate in an economy and culture that valued money over life. They were smart enough not to invent chainsaws and fellerbunchers (huge shears on wheels that roll along the ground, severing trees and stacking them into piles). They were smart enough not to invent wood chippers or pulp mills. They were smart enough not to invent an economy that ignored everything but cash. They were smart enough not to invent limited liability corporations. They didn't export mountains of timber overseas. They knew trees and other nonhumans as intelligent beings with precious lives worth considering, and not as cash on the stump, or resources to be managed, or even as resources at all. Their spiritual beliefs did not include commands to "subdue the earth," nor was their cosmology based on the absurd notion that one succeeds in life by outcompeting one's human and nonhuman neighbors.

And the Indians didn't subdue the earth. There is absolutely nothing in our culture's history to suggest that we can "equal or better this record of stewardship." There is everything in our culture's history and present practices to suggest that the deforestation will continue, no matter the rhetoric of those doing the deforestation, and that ecological collapse will be our downfall, as it has been for earlier civilizations.[7]

But believe neither us, nor even contemporary accounts of early explorers who wrote of the extraordinary richness of native forests, nor especially the handsomely paid liars of the timber industry and the government. For the truth lies not in what they say, nor even in what we say. The truth lies on the ground. Go out and walk the clearcuts for yourself. Rub the dried soil between your fingertips. Walk the dying streams; listen to the silence in the skies (except for the whine of chainsaws and roar of distant logging trucks). Walk among ancient ones still standing, trees sometimes two thousand years old. Put your hands on their bark, on their skin. Taste the difference in the air. Smell it. Reflect on the beauty of what's still there, and on what has been lost—what has been taken from us.

When you've finished crying, and if you want to know more about the current crisis in the forests—where we are, how we got here, and where we're going—then come back and read the rest of this book.

I walk in an ancient forest. Redwoods who sprouted long before civilization reached this continent surround me. When a redwood falls, young trees often come up from burls around the base or underground, so when you see several two-thousand-year-old trees huddled around a space that one day might have been another massive trunk, it's easy to find yourself slipping even further back in time, perhaps another two thousand years, to when that parent tree sprouted.

Many parts of this forest floor never see the sun. Big-bodied, small-headed beetles scurry beneath ferns that run like a carpet between trees. Hard-shelled millipedes wriggle through duff. At every step my feet encounter rough surfaces of redwood roots that twine together tree-to-tree to hold these trunks upright through storm and wind. The roots, I've read, seek each other out to form nets of mutual support. Would that we remember to do the same.

An old alder, downed in the last windstorm, cuts across the path. It died long before it fell, its branches growing bearded with moss. It served the forest when it grew, it served standing after death, and now it will serve the forest as it slowly falls apart.

I make my way to a large stream. I stand in the soft soil of a thousand years of fallen leaves, and look out to see a salmon sweeping clean her nest. Her body is big and dark brown, her tail white and tattered from the journey upstream and from beating against the gravel beneath her. A sudden sound pulls my attention downstream, and I see another fish fighting her way up a series of rapids. She makes it halfway, then tires, or maybe realizes she's chosen the wrong path, and floats back down. She rests a moment, then slides up again, sometimes shifting to her side—perhaps to keep as much of herself as possible under water, or perhaps to keep her belly from scraping too much on the bottom. She thrusts herself forward against the force of the water, this time heading directly for underwater paths she can most easily follow to the pool waiting above a final row of rocks. She swims for the only break in that row—revealing an extraordinary ability to read and analyze currents, to precisely predict upstream barriers by what she sees and smells and feels in the water moving around her—and makes it to the pool. She cruises quickly to the dark at the bottom, and I do not see her again.

When you consider the current landscape of the cradle of civilization—what is now Iraq and environs—what pictures come to mind? If you're like me, the images are of barren plains and even more barren hillsides, goats or sheep grazing on a few scrubby bushes breaking a monotony of light brown dirt. But it was not always so. As John Perlin states in *A Forest Journey: The Role of Wood in the Development of Civilization,* "That such vast tracts of timber grew near southern Mesopotamia might seem a flight of fancy considering the present barren condition of the land, but before the intrusion of civilizations an almost unbroken forest flourished in the hills and mountains surrounding the Fertile

Crescent."[8] The trees were cut to build the first great cities and the ships that plied the first great empire. Once the ships were built, wood was imported to make the cities even bigger. Down went the great cedar forests of what is now southwest Turkey, the great oak forests of the southeastern Arabian peninsula, and the great juniper, fir, and sycamore forests of what is now Syria.[9]

One of humanity's oldest written stories—one of the formative myths of Western culture—is that of Gilgamesh, who destroyed southern Mesopotamia's cedar forests to build a city.[10] According to this story, Enlil, the chief Sumerian deity, who must forever watch out for the well-being of the earth, entrusted the demigod Humbaba to defend the forest from invaders. But the warrior-king Gilgamesh killed Humbaba and leveled the forest. Enlil sent down curses on the deforesters: "May the food you eat be eaten by fire; may the water you drink be drunk by fire." These curses have followed us now for several thousand years.

Let's move a little west. Picture this time the hills of Israel and Lebanon. I recently asked a man from Israel if his country has trees, and he said, "Oh yes, we have lots of little trees, which we water by hand." This fits with the images that come to mind. Every picture I've seen of the Crucifixion, for example, shows a hilltop devoid of trees. The same is true for most of the pictures I've seen of Palestinian refugee camps, and for Israeli settlements. What happened to the "land of milk and honey" we read about in the Bible? And what about those famous "cedars of Lebanon?" You'll find them only on the Lebanese flag now. The rest are long gone—cut to build temples, cities, and ships, cut for fuel, cooking, metalworking, pottery kilns, and all the trinkets of commerce.

Move west again, to Crete, and then up to Greece, and we see the same stories of trees making way for civilization. Knossos was heavily forested and now is not. Pylos, the capital of Mycenaean Greece, was surrounded by giant pine forests. Melos became barren. The same is true for all of Greece.

When you think of Italy, do you think of dense forests? Italy was once forested. These forests fell beneath the axes of the Roman empire.

Or how about North Africa? Surely not. This land is as barren as the Middle East. But here, once again quoting Perlin, "Berbers fulfilled their duty by felling the dense forest growth for their Arab masters. Such large quantities of wood were shipped from these mountains that the local port was named 'Port of the Tree.'"[11] All to make Egyptian warships.

We could continue with this journey, through France and Britain, across North and South America, into Asia and Africa, but by now you see the pattern.

That pattern continues today, accelerating as our culture metastasizes across the globe. Worldwide, forests fall.

As of 1997, Nigeria had lost 99 percent of its native forests.[12] The same was true of Finland and India. China, Vietnam, Laos, Guatemala, Ivory Coast, Taiwan, Sweden, Bangladesh, the Central African Republic, the United States, Mexico, Argentina, Burma, New Zealand, Costa Rica, Cameroon, and Cambodia had all lost at least 90 percent. Australia, Brunei, Sri Lanka, Zaire, Malaysia, and Honduras had lost at least 80 percent. Russia, Indonesia, Nicaragua, Bhutan, and the Congo had lost at least 70 percent. Gabon, Papua New Guinea, Panama, Belize, Colombia, and Ecuador had lost at least 60 percent. Brazil and Bolivia had lost more than half. Chile, Peru, Canada, and Venezuela had lost almost half.

Since 1997, of course, things have gotten much, much worse.

Forest Dwellers

> We would never buy paper made from dead bears, otter, salmon and birds, from ruined native cultures, from destroyed species and destroyed lives, from ancient forests reduced to stumps and mud; but that's what we're buying when we buy paper made from old growth clear-cut trees.[1]
>
> *Margaret Atwood*

When a forest is cut, not only trees are killed. Whether it's lions in ancient Greece, spotted owls or coho salmon right now in the Pacific Northwest, or gorillas in Africa, the loss of forests means the loss of the creatures who live there.

The list of plants and animals damaged or extirpated by the deaths of once-great forests is long, and getting longer every day. Golden-crowned lemur, orangutan, Siberian tiger (of whom there are only two hundred and fifty left), marbled murrelet, Port Orford cedar (killed by a fungus transported on logging equipment), black forest wallaby, aye-aye, red cedar, mahogany, ivory-billed woodpecker, Carolina parakeet, golden-capped fruit bat, Hazel's forest frog, smooth-skinned forest frog, Amur tiger, Amur leopard, forest owlet, Nelson's spiny pocket mouse, saker falcon, red wolf, panda, and on and on.

Scientists estimate an average of 130 species are driven extinct every day. That's about fifty thousand each year. That is not just by deforestation, but by the larger effects of industrial civilization. Deforestation plays its part, though, in great measure because forests are home to so many creatures. For example, although rainforests presently cover only 3.5 percent of the planet's land surface, they support more than half of all known life forms. The national forests of the United States provide habitat for three thousand species of fish and wildlife.

Seventy-five percent of the mammals endangered by the activities of industrial civilization are threatened by loss of forest habitat.[2] For birds, the figure is 45 percent. For amphibians it's 55 percent, and for reptiles it's 65 percent.

Even those apologists for industrial forestry who admit other creatures besides humans live on this planet, and who acknowledge that destroying their homes could possibly harm them the tiniest little bit, still argue that logging is a trivial cause of damage compared to mining and agriculture. They especially like to show pictures of poor (brown) people using slash-and-burn agricultural techniques in the rainforests. But this argument is as much a deflection as most of their others. Worldwide, logging likely accounts for more than two-thirds of forest destruction, as opposed to burning and other causes.[3] In Oceania it's "only" 42 percent. Asia, 50 percent. Central America, 54 percent. South America, 69 percent. Africa, 79 percent. Europe, 80 percent. North America, 84 percent. Russia, 86 percent.

Recent studies show, too, that species extinction likely continues for a century after deforestation.[4] Guy Cowlishaw of the Zoological Society of London cautions, "We should not be lulled into a false sense of security when we see that many species have survived habitat loss in the short term. Many are not actually viable in the long term. These might be considered 'living dead.'" By correlating, for example, the number of individuals of different species of primates living in Africa, their habitat size, and the extent of deforestation of their habitat, he has come to the conclusion that deforestation is leaving Africa with a large extinction debt. Even if no additional forest is cut, six countries—Benin, Burundi, Cameroon, Ivory Coast, Kenya, and Nigeria—stand to lose more than a third of their primate species in the next thirty or forty years. That presumes, once again, *no further deforestation*. But scientists estimate that within that same time, 70 percent of remaining West African forests and 95 percent of remaining East African forests will be cut.

It's not just primates. Studies on birds show similar trends. Thomas Brooks, a biologist from the University of Arkansas who has studied avian extinction in Kenya's Kakamega Forest, said, "Even a century after a forest has been fragmented, it may still be suffering from bird extinctions. . . . The good news is we have a brief breathing space. Even after tropical forests are fragmented, there is still some time to adopt conservation measures to prevent the extinction of their species. The flip side of this is bad news, though: There is no room for complacency."

Healthy forests are crucial not only to the creatures who live there. Forests purify water and air. They mitigate global warming by storing carbon. Because half of the rain in rainforests comes from local water evapotranspirated from the forest itself, forests increase local precipitation. They prevent flooding and erosion.

It is common when making a plea to halt deforestation to talk about the ways the loss of these forests hurt *us,* using, for example, the fact that rainforests can be considered great medicine chests, if only we will use the medicines instead of destroying the chests. Just tonight I read on a website deploring tropical deforestation, "The rainforest is the earth's natural laboratory, from where one quarter of today's pharmaceuticals are derived. One seemingly insignificant plant, the rosy periwinkle, gave us medicines which revolutionized the treatment of leukemia in children. According to the National Cancer Institute, 70 percent of the plants used in fighting cancer can only be found in the rainforest. But less than one percent of tropical forest species have been thoroughly examined for their medicinal properties."[5]

While it's certainly true that there are many selfish reasons to stop cutting down forests, we don't want to emphasize them, because ultimately—and even in the short run—we don't think that particularly helps. It doesn't challenge the grotesquely narcissistic and inhuman utilitarian perspective that *is* our worldview and underlies our attempts to dominate the world.

A few years ago I was one of the only environmental representatives at a conference of children's health advocates. That in itself was strange, I thought. How can you possibly discuss the health of children without emphasizing the fact that industrial civilization is rendering the planet uninhabitable for them?

One of the advocates there—a high level federal bureaucrat at the Centers for Disease Control—expressed the need to halt tropical deforestation (it often seems to me that more people in the United States want to halt tropical deforestation than want to stop it here at home) by saying, "We need to save those plants because they're our medicines for the future."

"That's precisely the problem," I responded. "The belief that the forests belong to us. They're not *our* medicines, and they're not *our* forests. First, the plants belong to themselves, and they belong to the forest. Second, if they belong to any humans at all, they belong to the indigenous people who live on that land. We have no more right to take their plants for medicines than we do for timber."

Several people looked at me as though I had suddenly stopped speaking English and begun quacking like a duck. This is what often happens when you cease to speak the language of unbridled exploitation—untethered selfishness—and begin to suggest that forests, and the creatures who live in them (including indigenous humans), have the right to live on their own, regardless of how useful or not they may be to us. What was happening in that room was in many ways what happens moment-by-moment in the forests: a clash of incompatible worldviews and value systems.

At every step of the way there have been humans living in the forests that have fallen to the axe, and now the chainsaw: People who do not view forests as resources, but instead as homes to be lived in forever. There were the indigenous conquered near the Fertile Crescent, whose sacred groves were cut by Gilgamesh and his ilk. The Canaanites and many others, conquered in the Promised Land, whose sacred groves were cut by the Israelites lest the Israelites be tempted to worship in their shade. The indigenous of northern Greece, whose forests were cut to serve commerce, and who were called *barbarians* because they did not speak the language of civilization, but instead made sounds like *barbarbar*. These people were conquered, their forests cut. The indigenous of Italy, France, England, called *savages* because they lived in forests (*savage* derives from the root word for *forest: savage:* "not domesticated, untamed, lacking the restraints normal to civilized human beings," from Medieval Latin *salvaticus;* alteration of Latin *silvaticus,* of the woods, wild; from *silva,* wood, forest). These, too, were killed, their lands deforested.

Move across the ocean to the United States. A standard conceit of the settlers was that they faced not *terra incognita* but *terra vacuuis,* an empty land with trees ripe for cutting. But these were not empty lands, and they are not empty lands today. There are those who live there. There are nonhumans, whose lives are as meaningful to them as yours is to you and mine is to me. And there are humans, with lives just as precious.

Wilderness is a social construct. My niece recently moved to Louisiana, and sent me a note in which she stated how uncomfortable she is that an alligator lives on her Coast Guard base. "Call me crazy," she wrote, "But I think it's odd to have wild animals so close to where people are." Not always would this have seemed odd. For almost all of human existence, it was simply how things were. And for some humans it still is. For them there is no city in here, no wilderness out there, no split between humans who exploit and a resource base to be exploited.

What all of this means is that when we talk about saving

forests we too often forget about the people who call them their home. No, we're not talking about those people with more cash than integrity who buy ecologically sensitive pieces of ground and threaten to construct vacation homes—with the real purpose being to extort money from those who wish to protect the land. Nor are we talking about transnational timber corporations attempting to "gain access to" wild forests the world over. Nor are we talking about loggers, many of whom truly do love to walk in the forests they're destroying. Nor are we talking about environmentalists living in yurts and composting their feces into humanure. We're not *even* talking about writers and researchers who love to look at salmon and will do anything possible to help stop deforestation.

We're talking about the indigenous, those who live on the land that their ancestors lived and died on, going back so many generations that the distinction is lost between those who live on the land and the land itself. We're talking about those whom we have never gotten to know, and who have never fit our self-serving stereotype that they are "beastly," "savage," "primitive," somehow subhuman, living lives that are "nasty, brutish, and short." This notion is self-serving because it reinforces the conceit that these people would be better off if we civilize them, take them (by force if necessary) out of their childlike ways to live as adults. As Ronald Reagan put it, "Maybe we made a mistake in trying to maintain Indian cultures. Maybe we should not have humored them in that, wanting to stay in that primitive lifestyle. Maybe we should have said: No, come join us. Be citizens along with the rest of us."[6] Conveniently left unsaid is the theft of their land, and its ultimate despoliation.

Nor do the indigenous live romantic lives wandering about picking a few berries now and then. They have serious long-term relationships with the plants and animals with whom they share their landscape. Ray Rafael, who has written extensively on the concept of wilderness, has said, "Native Americans interacted with their environment on many levels. Fortunately, they did so in a sustainable way. They hunted, they gathered, and they fished using methods that would be sustainable over centuries and even millennia. They did not alter their environment beyond what could sustain them indefinitely. They did not farm, but they managed the environment. But it was different from the way that people try to manage it now, because they stayed in relationship with it."[7]

Theft of indigenous land is not ancient history, something that *only* happened a long time ago, something to express our regrets over as we continue to profit from their land. It happens today, all over the planet. Anywhere there are indigenous people living traditionally in forests, they are being threatened, harassed, arrested, dispossessed, killed, and their forests are being cut down. Here are a few current examples among far too many.

Africa: The Bayanga Wood Company deforests the homeland of the Ba'Aka (pygmies) of the Central African Republic. The Ba'Aka are forced into settlement camps at the fringes of their dying forests.[8] The transnational timber corporations Rougier (French), Danzer (German), Feldmeyer (German), Wonnemann (German), and the Dutch-Danish-German consortium Boplac deforest the Congo. Pan African Paper Mills, Raiply Timber, and Timsales Ltd. are entering—and destroying—the forests of the Ogiek people of Kenya, who are being evicted from where they have lived, hunted, and gathered honey forever. In 1967 the World Bank decided that the Gishwati forest, home to the Batwa (pygmies), should be cleared to use for potato farming and cattle raising. The Batwa were not, of course, consulted. As a sixty-one-year-old Batwa says, "We were chased out of our forest, which was our father because it provided us with food through gathering and hunting. . . . The State chased us out of the forest and we had to settle in the fringes, where we die of starvation. All the development projects that were carried out in Gishwati forest have done nothing for us and no Batwa has even received the benefit of a job."

The genocide continues. A 2002 news report (not from the corporate press, of course, but from the human rights organization Survival International) stated that the Botswana government "denied the Gana and Gwi Bushmen still in the Central Kalahari Game Reserve their only means of communication with the outside world, and turned back Bushmen bringing them essential supplies of food and water. Government officials seized solar powered radio transceivers, provided by [Survival International] for the Bushman communities. They also told two Bushmen bringing food and water to the beleaguered communities, whose supplies were cut off by the government last week, that entry to their ancestral lands was forbidden. The two were later allowed to deliver the food and water, but were told that in future they would have to have a special permit or pay to enter the reserve. The Central Kalahari Game Reserve was set up in the 1960s as a home for the Gana and Gwi Bushmen, whose ancestral lands include the reserve area. Yet since the mid-1980s, the Botswana government has waged a campaign of harassment to force the Bushmen off the land that is theirs under international law. In past weeks many of the 700 Bushmen still living in the reserve in the face of this harassment have been forced to leave, and last week the government terminated supplies of water and food to those who are still resisting."[9]

Back to the "developed" world. North America. British Columbia granted huge timber concessions to the timber giant Macmillan-Bloedel, which made billions of dollars by clearcutting nearly all of Vancouver Island. In 1999, Mac-Blo, as it is commonly known, was bought out by the U.S.-based transnational timber corporation Weyerhaeuser, which had already liquidated forests in Wisconsin, Minnesota, Washington, Oregon, the

Philippines, and Indonesia. Weyerhaeuser, like Mac-Blo before it, is clearcutting like mad, in part because the First Nations of Canada have never extinguished title to the forests being clearcut and are suing the Canadian government to exercise their rights to sovereignty over this land, including not allowing it to be cut. The Haida have sued Weyerhaeuser for illegally clearcutting their land in the Queen Charlotte Islands. Guujaaw, chief of the Haida in British Columbia, said about Weyerhaeuser, "They've come and wiped out one resource after another. . . . We've been watching the logging barges leave for years and years, and we have seen practically nothing for Haida."[10]

South America. The Guarani living in forests in Argentina do not believe land can belong to anyone. How can human beings, who are only passing through life, be owners? The Moconá S.A. Forestry Company, which is not a human being but a corporation, a legal fiction, is cutting down their forests. The company offered each community seventy-four acres on which to live. The Guarani rejected the possibility that the land could have any owner and found it absurd that they were being offered seventy-four acres of those communal lands where their ancestors had lived and where they themselves were already living, land they were, according to their worldview, borrowing from their children. The corporation raised the offer to about five hundred acres, and continues to cut.

The Wichí have lived on the same land (in what is now called Argentina) for at least 12,000 years; now through depredations of timber and agricultural corporations, their homeland has been reduced from more than 170,000 acres to less than sixty-seven. The remaining sixty-seven acres are an oasis of green amidst a now-barren landscape.

The Mapuche of Chile have lost more than 95 percent of their original 27 million acres, and now logging companies are coming for the rest. Police murder children who protest the logging.

Asia. The Karen in Burma are under attack from Canada's Ivanhoe Capital Corporation, which in 1994 reached an agreement with the Burmese military regime to run the Monywa copper mine. Safety measures are completely absent. Miners threaten to blow up local residents who complain about water pollution and skin problems. The Karen are also under attack by the United States' Unocal corporation, which along with the military has used forced labor to construct the Yadana gas pipeline. Mass murder and mass rapes are useful tools for enslaving a people and forcing them to destroy their own landbase. And the Karen are under attack by the Thai dam-building company GMS Power and the Electricity Generating Authority of Thailand, which are building a huge dam at the Salween River, the only remaining free-flowing major river in the area. One hundred and seventy-five villages will be relocated. Or perhaps not. The Burmese Army has begun a program of extermination.

The Togeans of Indonesia have taken to torching logging equipment of the transnational timber corporation destroying their home.

In the Philippines, logging companies and the military have taken over the forests of the Agta, who are now homeless and still menaced. A spokesperson for the Agta recently stated, "A certain colonel warned us that if we do not vacate our land, our tribe will be exterminated."

The Penan of Malaysia have been struggling for their lives and for the life of their forest for many years. But life was not always a struggle. As Ngot Laing, Chief of Long Lilim, Patah River, said, "In the past our life was peaceful, it was so easy to obtain food. You could even catch the fish using your bare hands—we only needed to look below the pebbles and rocks or in some hiding holes in the river." Urin Ajang concurred, "In the past, we did not fall sick, we did not have scabies, the water was clean. We did not have all these puddles that breed mosquitoes." But now, Ngot says, "The people are frequently sick. They are hungry. They develop all sorts of stomach pains. They suffer from headaches. Children will cry when they are hungry. Several people including children also suffer from skin diseases, caused by the polluted river. Upper Patah used to be so clean. Now the water is like Milo, sometimes you can even find oil spills floating downstream." Another Penan, Lep Selai, said, "Living a settled life is just not our way. We are used to the forest. Besides, I do not know how to farm." This doesn't mean the Penan are too stupid to become farmers. The real point is, as Peng Megut put it, "We know that if we agree to settle down, it would in effect be a tradeoff for our forest. The government is asking us to settle down, as if once when we are settled, they can do anything to our forest." Ayan Jelawing sums up, "We were the first people of this Apoh area. The waters did not have a name then, not until we gave it a name in our language. . . . The logging companies first entered into the Apoh area in the 1980s. When the Penan communities went to meet the companies' managers they would simply say that the Penan do not have any rights to this area. How could this be?" Ajang Kiew states, "We asked for forest reserves. We asked for a school for the village. We asked for clinics. Instead they gave us the logging companies. Now it is oil palm plantations. We would end up as laborers for hire. The profits would only make other people rich. But the land they work on is land belonging to the Penan." And finally, Nyagung Malin gives a solution: "We are used to living in the forest. And life did not used to be difficult. If we needed to build our huts, we could easily find the leaves in the forest. If you really want to give us development, then do not disturb our forest."[11]

The people of the forests aren't stupid, backward, or stubborn; they are loyal to the source of life.

Accountability

> We policemen have been made the tools of the big business interests who want to run things. I'm ashamed of myself for consenting to do their dirty work. The big fellows in this town can do anything they like and get away with it, but the workers can't even think what they want to think without being thrown in jail.[1]
>
> *Police Captain Plummer*

The military and police, and, more broadly, the government—any government—often promote deforestation, and spend far more time and energy working toward the theft of indigenous land than its protection.[2] This was true in the days of Gilgamesh's Mesopotamian city-state of Uruk, and in the days of the Israelites, and true in the days of the Greeks and Romans. It's been true throughout American history, and it's true today. This support is quite often direct, as when the military in Papua New Guinea machine-guns those who resist Freeport McMoRan's copper and gold mining there; as when the Saramake people of Suriname are threatened with imprisonment when they resist the deforestation of their land by Chinese timber companies; as when the Indonesian military suppresses those in the path of ExxonMobil's oil operations; as when police in the United States frequently use pepper spray and "pain compliance holds" against those who attempt to halt deforestation here.

The support comes, sometimes, through intentional neglect, and through repeatedly refusing to enforce any kind of accountability on those who deforest. Enforcement officers, politicians, bureaucrats, police, judges, and businessmen are tied together in patron-client networks that promote their own interests rather than enforcing the community's forest policies and laws.[3]

We want to tell you, for example, a story about the relationship between the government and the ongoing destruction of the last redwood forests in the United States. It concerns a timber company called Pacific Lumber (PL).[4] As recently as two decades ago, PL was a family-owned company known for being fair to its workers and for being as sustainable as an industrial forestry company can be (which isn't terribly sustainable, but one of the first lessons you learn as an environmentalist is to savor bright spots—or less gloomy spots—where you find them). Then the owners decided to take the company public.

The company was soon taken over by a corporate raider named Charles Hurwitz, famous for proclaiming and actualizing his version of the Golden Rule: He who has the gold rules. Hurwitz has a long history of illegal and antisocial activities, stretching back to his early twenties, when he was forced to plead no contest to the Securities and Exchange Commission for illegal stock market dealings. He later acknowledged looting New York-based Summit Insurance out of $400,000. Next he raided the pension fund of Simplicity Pattern, causing retirees' benefits to drop from $10,000 per year to $6,000 per year. That company, under the new name MAXXAM, became the holding company through which Hurwitz has raided many other companies, bilking retirees, stockholders, and the public out of money, breaking unions, and eventually, as we'll see, devastating the landscape of northern California. During the Savings and Loan scandal of the 1980s, Hurwitz and MAXXAM looted the United Savings Association of Texas, costing taxpayers $1.6 billion to bail out. More than $1 billion of this money remains unaccounted for, despite (or perhaps because of) lukewarm prosecution by the United States Department of Justice.

Hurwitz used some of his ill-gotten gains to take over Pacific Lumber in northern California. One of the first things he did was raid the workers' pension fund, taking $55 million from retired loggers and millworkers. Then he began liquidating the company's assets, including the world's largest stands of privately owned (well, actually corporate-claimed) old-growth redwoods. Simultaneously, his longtime partner in crime and number two man at MAXXAM, Barry Munitz, resigned to become Chancellor of the California State University System. They got a pet state senator, Barry Keene, to secure the passage of a resolution creating the Center for the Resolution of Environmental Disputes, at the head of which would be, you guessed it, the Chancellor of the California State University System. Moreover, PL donated $61,000 to California Governor Gray Davis. Davis then solicited and received a $15,000 contribution from MAXXAM for one of Davis's political pals right when California was considering regulatory action against the company for water quality violations. You may be familiar with this tactic under its street name: *shakedown*. It is often stated in California that Gray Davis is an honest politician, by which it's meant that when he's bought he stays bought. That's certainly true in this case. The North Coast Regional Water Quality Control Board, with members appointed by Governor Davis, has repeatedly deferred action on water quality matters pertaining to PL.

PL routinely breaks state and federal law. Even with regulatory agencies in its pocket, it's been cited hundreds of times for violations of forest practices rules, the Endangered Species Act, the Clean Water Act, and so on. Mudslides from PL clearcuts have destroyed (human) homes. They've destroyed water supplies for (human) communities. A few years ago, after a logger threatened to kill protesters ("Ohhhhh, fuck!" he is caught screaming on videotape, "I wish I had my fucking pistol! I guess I'm gonna just start packing that motherfucker in here, 'cause I can only be nice so fucking long!"), the logger actually did drop a tree on one of the protesters. The logger was never arrested. Indeed, Humboldt County sheriffs assaulted and arrested environmentalists instead, and the local district attorney issued an opinion that the environmentalists themselves should be charged with manslaughter.

Meanwhile, the cutting continues.

To much fanfare, and over the objections of local environmentalists, Dianne Feinstein, another of Hurwitz's pet senators—federal this time—pushed through a deal that gave Hurwitz $380 million in exchange for 7,500 acres of redwoods. Even more important to Hurwitz than the money, if such a thing is possible, was that as part of the bargain, the feds agreed to allow Hurwitz to deforest another 46,000 acres, including 2,000 acres of old growth, over the next ten years. Hurwitz will also be allowed to deforest much of the rest of the 200,000 acres claimed by PL, including 8,500 acres of old growth, over the next fifty to one hundred years. Further, the deal waived compliance with the Endangered Species Act in many areas, tantamount to giving PL a fifty-year permit to kill endangered species.

Local environmentalists sued over that deal, and in the three years since, PL and the government agencies that protect it have refused to hand over the applicable records to the court, presumably because of what they would reveal about the deal and the effects of PL's logging, were they to become public. Finally, a judge issued a stay on all PL logging associated with the deal until documentation was released.

The response by PL was, unsurprisingly, to ignore the stay and to continue—in fact to accelerate—logging. A further response by PL's president and CEO Robert Manne was to call those who opposed this logging "eco-terrorists," and to say that their actions fit "a pattern of behavior that the Department of Justice will be keenly interested in reviewing."[5] He continued by stressing that "ours is a society of laws and rules, and we are troubled that these activists are obviously determined to ignore both. This illegal and aggressive behavior must not be allowed to continue." I sometimes wonder the degree to which this sort of extreme irony is intentional, and the degree to which it is unconscious. If the former, he's evil. If the latter, he's stupid. We suspect a combination.

Here's the point: As PL loggers cut trees in explicit and knowing violation of the judge's stay, as well as any number of federal and state laws and regulations, they have been escorted by members of sheriffs' departments, not to make sure they don't continue their illegal cutting, but to make sure they can. On the other hand, since the stay, sixteen environmentalists have been arrested, many carrying copies of the judge's orders. As deputies carried one eighteen-year-old female tree-sitter from a logging site (after having put her in pain-compliance holds), they said to her, "We're good citizens. We remove trash from the forest." Her bail, because she was protesting illegal cutting by PL, has been set at $200,000.

Meanwhile the cutting continues.

We don't want to give the impression that PL is unique. Far from it. The political alliances, lax enforcement of forestry rules, and exemptions from the Endangered Species Act and other environmental regulations are all industry standard. For example, while Hurwitz has gotten a lot of bad press, Sierra Pacific Industries (SPI) has been, according to one researcher, Doug Bevington, "quietly plundering the state's forests on a scale that makes Charles Hurwitz look like a novice."[6] SPI claims 1.5 million acres of land in California, making it the state's largest private (read, corporate) landowner and the second-largest landowner in North America.

SPI benefits greatly from federal subsidies, first by cutting 39 percent of all the trees taken from the national forests in California, and then using those profits to double its corporate land holdings over the last ten years.

Between 1992 and 1999, SPI increased its clearcutting by more than 240 times, and increased the size of its average clearcut from

46 to 361 acres. The corporation has plans to clearcut a million more acres (1,562 square miles), an area larger than Rhode Island.

The California Department of Forestry is essentially owned by the timber industry, and yet SPI still routinely ignores CDF's rules. Between 1991 and 1999 Sierra Pacific asked the state to exempt 4.4 million acres from meaningful departmental oversight, leaving 711,445 acres where SPI *pretended* to abide by the regulations.

SPI's purchase of state regulatory agencies cost them $231,500 in political donations in 1998 and 1999. This money was used to purchase support at every level from the governor to county sheriffs, which might help explain who gets arrested and who does not. Although SPI purchased Governor Gray Davis's opposition in 1998, they have since shown themselves to be willing to purchase politicians of both major parties—revealing, if nothing else, that the parties are two sides of the same corporate coin. In 2002 they gave Davis $42,716, and more recently hosted a timber-industry fundraiser that brought in $129,000 for him. That may have been a belated payment for Davis's 1999 appointment of Mark Bosetti of Sierra Pacific to sit on the State Board of Forestry. A spokesperson for Davis stated, "There's absolutely no nexus" between these pay-offs and Davis's actions.[7]

Not content with the purchase of taxpayer-subsidized timber, SPI has routinely stolen public timber from national and state forest lands as well. As federal agents wrote in a 1993 briefing paper, "These companies are suspected of manipulating the log scale [in order to escape paying full price for the timber they removed from public lands]. It is further suspected that various Forest Service officials have met with representatives of these companies concerning the administration of Forest Service timber sales and decisions have been made not in the best interest of the government. It is suspected those decisions have resulted in an undetermined loss of revenue to the United States Treasury." That case was mysteriously dropped. The company has been known, when cutting trees on SPI-claimed land, to cross borders into public lands and cut there as well. Or sometimes when cutting they'll simply sneak extra log trucks out without proper receipts. When individuals do this, it's called theft.

Not content even with all this, a few years ago SPI came up with something called the Quincy Library Group. The claimed purpose of the group was to deal with the (logging-induced) collapse of the regional forest-based economy. The group's name makes it appear to consist of a few folks hanging out in the public library to wrestle with the fate of their local forests, but the name is another beauty strip. The truth is that twenty of the thirty participants in this self-described "community group" were associated with the timber industry. Most of them were SPI employees. Not surprisingly, the group's plan was to double logging in parts of three national forests, and cut a network of long, thin clearcuts through the region, all under the pretense of forest health. Also not surprisingly, SPI was able to get Congress to attach a rider making this plan the law of the land. The Plan is expected to cost taxpayers at least $70 million in lost revenues. SPI will be the main recipient of this largesse.

Meanwhile the cutting continues.

We don't want to give the impression that SPI is unique. Far from it. This is industry standard. For example, a few years ago Weyerhaeuser was caught stealing trees from federal lands. Lots of them. Back in 1991 Congress had set up the U.S. Forest Service Timber Theft Investigations Branch (TTIB), in order to provide the illusion of doing something about the annual theft of up to a hundred million dollars of federal timber by timber corporations. This amounts to up to 10 percent of the total federal timber cut—and this figure comes not from environmentalists, but from a former Forest Service Chief, in other words, from someone firmly in the pocket of the timber corporations. As reporter Mike

Romano noted, "Nobody expected the task force to do much more than inoculate the Forest Service against critics, but it won a string of stunning convictions, including a record-setting $3.2 million case against the Columbia River Scaling Bureau in 1993. Later that year, [Mike] Marion's 10-man team launched a law enforcement initiative unprecedented at the Forest Service—three concurrent investigations into allegations of million-dollar timber theft, accounting fraud, and obstruction of justice by Forest Service field managers." One of the companies investigated was Weyerhaeuser. Marion and his staff found that Weyerhaeuser had illegally cut up to 6 million board feet of timber. They also suspected that Weyerhaeuser was illegally exporting trees cut from federal land. A report by the task force states, "In summary, the Government was giving Government timber away for free." The particular thefts being investigated had cost taxpayers more than $3 million.

Once it became known that Weyerhaeuser was under investigation, and could be prosecuted, Forest Service employees granted Weyerhaeuser retroactive permission to cut trees outside of its contracts. As one employee later put it in a sworn affidavit, "We felt we need to protect [Weyerhaeuser] from violating the contract." Further, Forest Service employees tipped off Weyerhaeuser to the undercover task force investigation and destroyed pertinent Weyerhaeuser files two days prior to their intended seizure by investigators. Why? One Forest Service supervisor explained that he exposed the covert probe "because he couldn't afford to jeopardize his good working relationship with Weyerhaeuser."

Despite the best efforts of many within the Forest Service to sabotage the investigation, an independent Forest Service review concluded "the probability of conviction is good, as is the probability of civil recovery."

Something had to be done. But how was Weyerhaeuser to be protected? As always, the answer was simple. In this case, Jack Ward Thomas, Forest Service Chief, abruptly disbanded the

TTIB. A plausible case has been made that the order to disband came directly from the White House. Clinton did not want to upset Weyerhaeuser, a Fortune 500 corporation valued at more than $8 billion at the time—worth $18 billion by 2002—and a corporate ally from his days as governor of Arkansas.

Regardless of who gave the order to shut down the investigation, the process itself was dirty. Assistant U.S. Attorney Jeff Kent, responsible for timber theft cases in the Northwest, wrote the Office of Inspector General, which has oversight over the Forest Service: "Even as Chief of Special Prosecutions in Chicago, responsible for corruption and organized crime cases, I have never encountered in my 20 years as a prosecutor such a concerted effort by management to impede and sabotage the Congressionally mandated mission [of a group like the task force] or such Machiavellian maneuvers."[8]

Meanwhile the cutting continues.

I want to share another story about accountability. I teach creative writing at a prison in Northern California. Last night I happened to ask one of my students why he's in prison. He laughed, embarrassed, then said, "I was high on crank and stole a videotape from a store."

"How long did you get?"

"Two years."

"I gotta know," I said. "What movie was it?"

He laughed again, then said, "*The Lion King*. I got two years for *The Lion King*."

I had another student serving a life term for stealing a bicycle from someone's garage, and another serving a life term also for stealing a videotape. I do not know what the movie was, but to have gotten that sentence, it surely must have been a lot better than *The Lion King*.

Killing Forests

> The rapidity at which this country is being stripped of its forests must alarm every thinking man. . . . It is high time that we should turn our earnest attention to this subject which so seriously concerns our national prosperity.
>
> *U.S. Secretary of the Interior Carl Schurz, 1877*

Killing trees is not the only way industrial forestry destroys forests. In fact some other parts of the process may be even more damaging. Road building, for example, is often more destructive than the clearcutting that follows. Logging roads are the primary source of soil erosion and landslides in disturbed forests. Sediment discharge from a logging road can be ten times greater than from the clearcut surrounding the road. And a clearcut is (eventually, hopefully) replanted, whereas a road lasts for decades. Roads alter water runoff patterns and permanently disrupt subsurface water flow. Heavy rains can plug culverts with debris, resulting in overflow that blows out the whole road, filling streambeds below with mud. Road sidecastings become saturated and give out.

All this sedimentation results in flooding, erosion, landslides, lowered water quality in streams, scoured and destabilized streambeds, and property damage.[1] Fish suffocate. Streams die. Studies have shown that logging roads trigger debris avalanches that accelerate erosion 25 to 340 times beyond that in intact forests.[2]

Further, logging roads (and adjoining clearcuts) are barriers to many creatures, such as woodrats, voles, mice, frogs, and beetles. When isolated populations are unable to breed, fragmentation contributes directly to extinction.[3] When adjoining areas are fragmented, some populations of large mammals such as grizzly bears have declined even in million-acre national parks.[4]

Roads also harm forests by contributing to the invasion of noxious species. These species can be microscopic, as in the root rot that is killing Port Orford cedars. It is transmitted from area to area in mud clinging to logging equipment and other vehicles traveling logging roads. The noxious invasive species can be somewhat larger, as in the scotch broom, Himalayan blackberry, and pampas grass that now blanket much of the Northwest and choke out native plants. And the invasive species can be larger still, as in illegal (and slob) hunters who cruise logging roads to poach critters who would otherwise be safe.

These aren't the only problems caused by roads. Roads can result in soil becoming poisoned with heavy metals, such as lead from leaded gasoline and lead oxide from tires.[5] Study after study has shown that plants next to roads contain higher concentrations of lead, and that increases in traffic volume correspond to increases in concentration. This lead then moves through the food cycle, leaving animals with reproductive impairment, renal abnormalities, and increased mortality rates. Further, lead concentrations in tadpoles living near highways can be high enough to cause physiological and reproductive impairment in the birds and mammals who eat them. Earthworms, too, can accumulate lead, as well as zinc, cadmium, and nickel (motor oil and tires contain zinc and cadmium, while motor oil and gasoline contain nickel), sometimes in high enough concentrations to kill animals that eat these worms.

By encouraging human access, roads also increase the number and magnitude of forest fires, although you'll never hear this admission from politicians and other propagandists for the timber industry who want to use the threat of forest fires to justify further deforestation.

Another way logging destroys forests is through the logging

itself. This begins with the machinery. Heavy machines—bulldozers, trucks, skidders, fellerbunchers—compact forest soil which prevents tree roots from spreading by eliminating the air space for the oxygen roots need. It increases the susceptibility of roots to pathogens. Soil compaction can last for decades, reducing the productivity of the entire site.[6] The "site preparation" that takes place after industrial logging also destroys soil structure and chemistry.[7] Soil scarification exposes to air and rain soils that have for centuries or millennia been held safely beneath leaf litter. The piling and burning of waste (called "slash" by the industry) destroys much of what remains of the forest's understory; exposes, compacts, and even sterilizes soil; removes the protective soil cover from the underlying roots, seeds, and spores; and volatilizes soil nutrients such as nitrogen and sulfur.[8] Bulldozing the burned piles deprives much of the area of the nutrients that remained in the ash. Soils that have been burned are less permeable to water, which is bad for plants and for the streams that receive excessive runoff. And slash burning is a common cause of forest fires.

Clearcutting, which is industrial foresters' favored "treatment" (and yes, they actually use the word *treatment,* though thankfully they do not preface it with *special*), more or less permanently eliminates natural forest structures. Natural forests are layered, with multiple canopies, small openings where the sun shines through, and darkened hollows where it does not. Different plants and animals thrive in the different habitats. And the habitats change over time. An old tree falls in a windstorm, taking out the younger trees in its path. Thus a small opening is born. Sun-loving plants fill in the opening. They grow larger, and over time begin to shade out those below, who are replaced by plants who live in the cool and dark. Soon, the new canopy is intact, but still lower than the older trees, creating a variegated texture high above, where humans do not see. And somewhere else an old tree falls.

Or maybe it doesn't. Politicians and other timber industry propagandists like to talk about dead trees "going to waste" in the forest unless they're cut and removed, but dead trees are at least as important to the health of the forest as are live trees. Standing dead trees, called snags, provide homes for birds who carve nests out of the softening wood. Squirrels, too, live there, as do many other mammals, birds, insects, amphibians, and reptiles. Fungi love standing dead trees, and many creatures love to eat fungi. When these trees fall, they remain habitat, though now for different creatures. They store water, and as they rot, they serve as "nurse trees" to provide nutrients for the next generation.

Clearcutting destroys that complexity, replacing it with bare ground. Even if replanting succeeds, and often it doesn't, the result is a single-age, single-height, often single-species plantation.

Another problem with clearcutting is that with the overstory gone, snow falls on bare ground. Normally when snow falls on a forest, some stays on trees, some settles on the ground below. When spring rains fall on this forest, the drops are caught and cushioned by the snow still on the trees. Chilled water drips slowly to the forest floor, if it even makes it this far before freezing again. Water from the snow on the ground that does melt is absorbed in duff, loose soil, and dead trees, then released slowly, over time, to make its way eventually to streams and rivers, or down into the aquifer. After deforestation, rain falls on snow covering bare ground, quickly melting the snow, which runs off in sheets, carrying soil with it. This can cause one-hundred-year floods—floods that should happen only once a century—to become annual or even semi-annual events. In the language of foresters, this is called a "rain-on-snow event." In the language of the rest of us, this is called a disaster. Timber industry propagandists do not speak of this at all.

Cutting trees near streams reduces shade, increasing water temperature. This kills amphibians and fish such as trout and salmon.[9] For at least the last decade it has been common knowledge among people paying any attention whatsoever to the forests of the Northwest that at least two hundred distinct runs of salmon are already extinct or at risk of extinction because of habitat loss and degraded water quality caused by logging, agriculture, and hydroelectric dams.[10] Even those who care for nothing but money—Listen up, politicians!—should care about the death of the salmon, because they are commercially valuable. It won't be long before we have to shift that verb to past tense.

Clearcuts and roads also destroy forests by creating "edge effects." Edges of forests are different from interiors. Some creatures prefer edges, some interiors. Many of those who prefer interiors are unable to live at the edges.[11] Punching in roads and cutting patches of forests create more of these transition zones from bare ground to forest. The Forest Service and other timber industry supporters often claim that their "treatment" improves wildlife habitat. This is sneaky on their part, since they're defining wildlife to be only those creatures who prefer edges, such as white-tailed deer. It is absurd, however, to use as a justification for deforestation the creation of more white-tailed deer habitat when white-tailed deer are overrunning many parts of the country (precisely because of deforestation, as well as widespread eradication of predators). A word of advice: When politicians, Forest Service bureaucrats, or timber industry spokespeople (all of whom have the same fundamental goal in mind: "get out the cut") say they're working to improve wildlife habitat, always, always, *always* ask, "Whose habitat improves? Whose habitat is degraded?" And then, of course, you should *always* expect them to lie to you. But you knew that already.

These edges so beloved by the Forest Service, Weyerhaeuser, and white-tailed deer are detrimental when they come to dominate the forests, as they have. Trees formerly sheltered are for the first time exposed to heavy winds. They often blow over, pushing the forest's outer edge into what used to be the interior. Even if they do not blow over, the insulating forest canopy is now missing, leading to greater temperature variation. The sun now penetrates to the forest floor, drying it out, killing the soil. Nonforest species invade from these edges, preying on and competing with forest species. For example, cowbirds—called buffalo birds until the eradication of the buffalo—parasitize the nests of other birds.[12] Deer browse the edges of these stands, wiping out seedlings and shrubs, moving the edge ever inward.[13] Doubtless you've heard of spotted owls. One of the reasons they're in trouble is that they're being pushed out of their homes by non-native barred owls who move in from the edges.[14] Interior forests are disappearing in the patchwork of clearcuts that now cover much of this country. A study in the late 1980s showed that 41 percent of the ancient forest surviving in the Olympic National Forest in Washington was less than six hundred feet from an edge, rendering it unsuitable habitat for interior-dwelling species.[15] Since then things have only gotten worse.

Dozens of species of plants, birds, amphibians, mammals, and insects—spotted owls, marbled murrelets, and red tree voles, to name just three—can survive only within protected stands of ancient forest. Dozens more need ancient forest for some portion of their life cycle, such as nesting, breeding, or feeding. These species are important to the healthy functioning of the entire forest community. For example, lungwort lichens fix nitrogen in the soil necessary for plant growth. Flying squirrels eat and spread mycorrhizal fungi crucial to the Douglas fir's ability to absorb nutrients and water.[16]

Seventy-five percent of the world's large intact forests in the temperate and tropical regions are now threatened. Between 50 and 90 percent of all land species on earth inhabit the world's forests.[17] The forests are being killed. What are you going to do about it?

• • • •

Once forests have been injured through cutting, they're often insulted by being poisoned. According to the 1998 Report of the U.S. Forest Service, 200,357 acres of the national forests were treated with 179,240 pounds of herbicides, algicides, plant growth regulators, insecticides, acaricides, pheromones, predicides, piscicides, repellants, and rodenticides.[18] These include such chemicals as ozone-depleting methyl bromide, chloropicrin (deadly to humans at concentrations of four to ten parts per million, it was used as a nerve gas in World War I; its main use now is on the home front as a biocide), carbaryl, diazinon, and strychnine. Pesticide use is also routine on corporate-controlled lands. Walk the mountainous moonscapes on timber company land: cut, burned, poisoned, sterilized, and blowing away.

The whole of the damage caused by the timber industry is greater than the sum of its injuries. The harm accumulates. Imagine someone cutting a pound of flesh from your left thigh. That not only hurts like hell, but now your body has to deal with the wound. Just as you begin to recover, however, the same people take a pound of flesh from your right thigh. *You can live with that, can't you? You've got plenty of flesh. One pound won't hurt.* Then they take a pound of flesh from your upper arm. One from your lower back. One from your abdomen. Each time, those doing the cutting point out that you've not yet died from these ever-so-slight losses—*a pound is less than one percent of your body weight, so quit your crying and let me cut on you, you big baby*—so it must not have hurt you. *You look good,* they say. *Better than ever*. And then they take a pound from your left calf. Next your tormenters do not allow you to sleep. They cut at you incessantly. They drain pint after pint of blood (*Mosquitoes drain blood,* they say, *and you survive that natural process, so this process must be natural, too!*) They put strychnine in your food. And then they cut. And they cut. And they cut. Your flesh becomes a mosaic, a patchwork of scars.

What happens next?

You die, of course.

The same is true of forests. Building roads causes a certain amount of damage. The use of heavy machinery causes even more. More by cutting trees, more by dragging them out of the forest. More by burning leftover brush and applying poisons to the landscape. Step by step they injure forests. Step by step the forests die.

It's all even worse than this. When timber corporations do replant, the new "forest" is usually a monocrop. It's no longer a forest at all, but a fiber plantation. Most forest creatures cannot live there any better than they could in an Iowa cornfield, but spruce budworms love monocrops of spruce trees, and mountain pine beetles love monocrops of pine trees. Fiber plantations are far more susceptible than forests to fires, diseases, and catastrophic outbreaks of insects such as spruce budworms, mountain pine beetles, and tussock moths.[19] Of course fires, diseases, and insect outbreaks are expected, predictable, and *beneficial* natural responses to monocrops, beneficial because by helping to destroy monocrops they prepare the long slow way for the return of diversity. And of course fires, diseases, and insect outbreaks prompt calls from politicians, Forest Service bureaucrats, and timber industry spokespeople for increased cutting, not only on the affected plantations, but most especially on all of the natural forests they can get their chainsaws on.

It's all crazy. And it's all killing the forests.

Even if we look at forests solely on industrial forestry's own terms, ignoring their beauty and sheer being, ignoring the *lives* of all the plants and animals that make up a forest and the life of the forest itself, ignoring all the humans whose livelihoods are destroyed by diminished forests and those indigenous peoples who have lived in these particular forests forever—that is, just taking into account the production of fiber—industrial forestry *still* damages forests. Chris Maser, former scientist for the Bureau of Land Management, quit because he could no longer support BLM destruction of wild places. He says, "I know of no nation and no people that have maintained, on a sustainable basis, plantation-managed trees beyond three rotations. The famous Black Forest in Europe is a plantation; it and other forests are dying at the end of the third rotation. The eastern pine plantations are dying. It's the end of their third rotation. We do not have any third rotations here. I was quoted on this in a newspaper article, a little out of context, but not badly. One of the gentlemen from industry said, 'Well, geez, we're in our third rotation and trees are growing better than ever.' But he counted nature's old-growth as one of those rotations. We're only now cutting the second rotation, and the forest is not producing as it did. We do not value the land if we harvest the land's products to a maximum and make payments in minimums. We spend the least amount possible on every acre and harvest the maximum amount possible. We are not in any sense willing to reinvest in any natural, renewable resource."[20]

There exist, of course, scientists who are willing (and eager) to say that plantations can live past the third rotation. But we've never met an employed scientist who was not fully aware who cuts the paychecks. The scientists who praise plantations generally work for big timber corporations, industry organizations, the Forest Service, or university forestry departments, which exist to serve the timber industry.

In the real world, plantation after plantation not only loses productivity but dies in its first, second, or third rotation. Here's a story of a plantation in the Amazon: In the 1970s shipping mogul Daniel Ludwig believed that there would be a worldwide shortage of wood pulp, which he could remedy by growing pulpwood. Following in the footsteps of Henry Ford, who had tried to build a rubber plantation in the Amazon in the 1920s, Ludwig bought 600,000 acres at the mouth of the Amazon River. In 1978 Ludwig had a pulp mill towed from Japan, to save the time of building it in Brazil. Despite evidence that trees grew better on soils with intact topsoil and organic matter, he bulldozed and burned the moist tropical forest and planted gmelina trees native to Asia. Problems began: The biomass of the gmelina was a quarter that of the native forest, leaf-cutting ants attacked the exotic pines, and fungus slowed tree growth and reduced the quality of the wood. Ludwig cut the gmelina every three years to avoid fungus, but this robbed the soil of nutrients. Ludwig tried planting Caribbean pine, and then eucalyptus. Effluent from the pulp mill (and fertilizer and agrochemicals from affiliated farming operations) were dumped into the Jari River, killing fish downstream. Ludwig spent about a billion dollars trying to establish rice farming, cattle and water buffalo, a kaolin mine, and building a company town with three thousand housing units, stores, schools, 6,500 miles of roads, a railroad, and an airport. But the workers brought in from poverty-stricken northwestern states kept leaving (200 to 300 percent turnover per year), and the trees failed to grow. Ludwig let the plantation go bankrupt, leaving the Banco do Brasil with $180 million in debt; the place was sold to a consortium of Brazilian banks and companies for $280 million, some of which was Brazilian public funds. Local and national agencies had to take over the funding and operation of the town. By the mid-1990s, the plantation (now called Companhia Florestal Monte Dourado) was growing a few trees and selling a little pulp, some of which was from native trees, to supplement the still-faltering exotics.[21]

Indonesia provides another case study of the effects of tree plantations in the real world. In 1998 fires burned more than 12 million acres in East Kalimantan. The fiscal cost of the fires to Indonesia was more than $9 billion. Most of the fire burned on plantations—almost two-thirds of the pulp wood plantations in

East Kalimantan were destroyed by the fires—and on land converted to agricultural use and then left fallow. Fire damage was by far the worst in areas that had been recently logged. Over 80 percent of the fires were caused by private companies belonging to powerful conglomerates with close connections to the U.S.-backed dictator Suharto and his family. No government action was taken against 176 companies named by the Forestry Minister as responsible for setting fires.

At the same time, less than one million acres of protected native forests burned.

None of this was unexpected. Tropical rain forests usually don't burn, because fuel loads are low, the trees and bushes are not highly flammable, and even during droughts humidity is high. But logging as well as slash-and-burn agriculture changed all that. The cut forests of Indonesia burned following droughts in 1982–83, 1987, 1991, and 1994, in great measure because logging waste and the dense undergrowth of fast-growing pioneer species provide tremendous amounts of fuel to feed huge forest fires.[22]

The World Bank (the policies of which encourage industrial logging and the exportation of wood) predicts that if current trends continue, lowland rain forests will become extinct in Sumatra by 2005 and in Kalimantan soon after 2010.

There isn't much chance these trends will change. International financial institutions such as the World Bank pressure Indonesia to increase exports of timber, paper pulp, and palm oil, despite the fact that over 60 percent of Indonesian timber comes from illegal logging. But as we saw in the United States case studies, the law is rarely a hindrance when it comes to getting out the cut.

Since the late 1980s, $12 billion has been poured into Indonesia's pulp and paper industry.[23] Even funding that is supposed to go for life-serving ends often doesn't. An internal World Bank report in 1998 estimated that a third of its project loans simply disappeared.[24] As much as 70 percent of the Social Safety Net funding for the poorest of the poor never reaches them, or even comes close. More than 5 billion dollars was funneled through the Reforestation Fund over the past decade, but the vast majority of this money was instead used to clear natural forests for plantations, to try to turn wetlands into rice farms, and to build paper pulp plants. Reforestation Fund money was also used to support Indonesia's state aircraft company, national car project, and currency.

Sumatra's Leuser region alone contains more than 25,000 of the world's known species, including 4 percent of all known bird species and 3 percent of all known mammal species on earth. Say goodbye to the orangutan, the Sumatran rhino and tiger and elephant, and many more.[25]

Pulping the World

> I do not believe that there is either a moral or any other claim upon me to postpone the use of what nature has given me, so that the next generation or generations yet unborn may have an opportunity to get what I myself ought to get.[1]
>
> *U.S. Senator Henry M. Keller of Colorado, 1909*

Logging harms not only forests, but other places as well. The industrial processes of lumber manufacturing, wood processing and preservation, and paper manufacturing are all toxic.

Before it is used in construction, wood must be dried. This process causes air pollution, plain and simple. But drying is far less toxic than other wood manufacturing processes. Wood is often chemically treated to keep it from rotting, pressure-treated with creosote, pentachlorophenol, or arsenic, and sometimes with copper napthenate, zinc napthenate, and tributyltin oxide, all dangerous chemicals.[2]

Engineered wood products such as fiberboard, particleboard, plywood, and I-beams also involve nasty chemicals, including volatile organic compounds, phenols, and formaldehyde.[3]

Let's focus for a second on formaldehyde. Wood panels such as plywood and particleboard are glued together with highly toxic urea-formaldehyde, which is later off-gassed into the houses and offices constructed with these wood panels. Formaldehyde is a colorless, strong-smelling gas normally present at levels less than 0.06 parts per million (PPM). Levels at or above 0.1 PPM can cause watery, burning eyes; burning in the nose and throat; nausea; coughing; chest tightness; wheezing; skin rashes; and so on. Levels above 2 to 3 PPM can cause bronchitis, fluid in the lungs, inflammation of the lungs and respiratory tract, pneumonia, and respiratory failure resulting in death. It is a known carcinogen. It is also a very common industrial and commercial chemical, with 1991 U.S. production at about 6.6 billion pounds.

The production of paper products is extraordinarily toxic. A simple sniff of the air in a town with a paper mill will tell you that. Every year, 150 chemical pulp mills in the United States release millions of pounds of toxic chemicals, including aluminum salts, acetone, ammonia, chlorine dioxide, chlorine, chloroform, dimethyl disulfide, dimethyl sulfide, hydrochloric acid, methanol, methyl mercaptan, nitrogen oxide, sulfur dioxide, sulfuric acid, formaldehyde, dichloromethane, tetrachloroethylene, chromium, and friable asbestos. The paper industry is among the top producers of toxic chemicals, though the industry's ranking has dropped since the Environmental Protection Agency took some of the chemicals off its list of pollutants that need to be disclosed by manufacturers. And the release of some of the most toxic chemicals, such as dioxins and furans, have never been disclosed.[4]

After the wood fiber has been pulped, it's often bleached with chlorine. The chemical industry itself is the largest user of chlorine, but the paper industry is number two. A hundred to a hundred fifty pounds of chlorine are used to produce a ton of paper.[5] What's wrong with chlorine? When organic matter (the wood) is combined with chlorine, you get chlorinated organic compounds—up to a thousand different ones in chlorine-bleached pulp plant effluent. Only a third of these compounds have even been identified, much less tested for toxicity, but some of those that have been tested prove to be among the most toxic chemicals humans have yet produced: tetrachloride, chloroform, chlorophenols, polychlorinated dibenzo-p-dioxins (PCDDs), polychlorinated dibenzofurans (PCDFs), and polychlorinated phenolic compounds (PCPCs). Other pulping byproducts include synthetic chlorinated hydrocarbons such as the polychlorinated biphenyls (PCBs)—which have been banned along with DDT, toxaphene, and chlordane.[6]

The pulp and paper industry is the second largest source of dioxin—after the burning of plastics, which produces by far the most. Dioxin is considered by some to be either the most toxic substance created by our industrial economy or close behind some substances created by the nuclear energy and nuclear weapons industries. There are others who suggest it shouldn't be in the top ten—which, given our culture's predilection for manufacturing toxins, would still be akin to entering the Baseball Hall of Fame unanimously on the first ballot—but everyone this side of a few industry stooges agrees it's dangerous stuff. The EPA has linked dioxin and related toxins to increased rates of breast, testicular, and prostate cancers, as well as infertility, impaired immune systems, and nervous system disorders.[7]

Of course nonhumans who live downstream of paper mills suffer, too. Dioxin from paper mills causes tumors in fish, deformities such as crossed bills in birds, and thinned eggshells and many other reproductive problems. Because dioxin bioaccumulates—is retained in the bodies of those who take it in, and then concentrated in the bodies of those who eat them, and so on, through the food cycle—predators suffer higher concentrations of this toxin. This includes humans: Because of dioxin contamination, those living downstream of mills are frequently advised not to eat fish caught in those waters.[8]

Further, the production of paper uses water, lots of it. The paper industry is the fifth largest industrial user of water in the world. Here in the United States the pulp and paper industry ranks number one.[9] When the water is flushed away into rivers, sewers, and lakes, it's full of chemicals.

After the wood or paper has been used, it continues to damage humans and the natural world, and not just because people read the corporate propaganda printed on it. The disposal of wood and paper products is a huge waste problem.

Here are some fun facts about waste: In 1988 the United States generated 180 million tons of municipal waste. If current trends continue, by 2010 the 275 million people in the United States will be producing 250 million tons of trash every year.

About a third of U.S. waste is generated by businesses. Office workers discarded more than 7 million tons of office paper throughout the country in 1988. Paper—including cardboard and newspaper—was the single largest component of the municipal waste stream: 72 million tons, or 40 percent of total waste. Office paper was the third largest category of paper waste, after corrugated boxes and newspapers. Office paper increased from 1.7 percent of the waste stream in 1960 to 4.1 percent in 1988, and is projected to be 6.4 percent by 2010, making it one of the fastest-growing categories of waste, despite e-mail and other technologies touted to save paper.[10]

People in the United States consume about 700 pounds of paper per person per year. Some states and demographic brackets are higher, with wealthier and more highly educated folks in California averaging 900 pounds per year.[11]

The EPA estimated that in 1990, 21 million tons of wastepaper—29 percent of the total consumed—was recovered for recycling.[12] The good news is that recycling numbers are going up. Here are some waste paper recovery rates, as of 1992—just two years later—by country: Taiwan, 56 percent; Netherlands, 53 percent; Japan, 52 percent; Germany, 51 percent; Sweden, 44 percent; United States, 39 percent; Denmark, 37 percent; Mexico, 35 percent; Canada, 35 percent; United Kingdom, 32 percent; Finland 28 percent; China, 24 percent.[13]

By 1997 the United States was recovering 45 percent of its waste paper, and recovered paper accounted for 38 percent of the industry's fiber needs (compared to 25 percent of fiber needs in 1987).[14]

In the United States, every 10 percent of recovered waste paper saves a million acres of forest from being cut.[15]

Bodyguard of Lies

> We've been lax in telling our story, and I think we've got a big job in front of us.
>
> *George Weyerhaeuser, in a speech to the Seattle Rotary Club, October 10, 1990*

We are often told that trees are "a renewable resource." Living most of my life in the West, much of it in areas under thrall to big timber corporations, I cannot tell you how many bumper stickers I've seen proclaiming that trees *are* America's renewable resource. Usually, these are next to stickers saying, "Get in, sit down, shut up, and hang on," "My wife tells me I don't listen, or something like that," or, incomprehensibly, "Save a tree, hug a logger." There's little evidence that forest soils—and therefore trees—are renewable beyond three rotations of cutting and removing, but there's a deeper point to be made here, having to do with the necessity of propaganda in order to perpetrate violence.

Winston Churchill famously wrote, "In war-time, truth is so precious that she should always be attended by a bodyguard of lies." I think Churchill, no stranger to lies both in war and peace, was on to something. But I would modify his statement to make it applicable to a war far more horrific and destructive even than World Wars I or II: our culture's war on the natural world. In this war, I would say that the truth is so *awful* that it must be attended by a bodyguard of lies. These lies can be callously and shrewdly manipulative, fabricated to pacify a public that might be raised to outrage were it not so easily and eagerly misled. Or sometimes the lies are believed by the perpetrators themselves. As Robert Jay Lifton made abundantly clear in his crucial book *The Nazi Doctors,* it is generally not possible to commit a mass atrocity without first convincing yourself—and not coincidentally others—that what you're doing is not harmful but instead beneficial. We're not killing forests, we're making toilet paper. *We're not killing Jews, we're purifying the Aryan race.* We're not killing forests, we're creating jobs. *We're not invading Russia, we're gaining lebensraum.* We're not killing forests, we're preventing wildfires. *We're not killing Indians and stealing their land, we're fulfilling our manifest destiny to overspread the continent.* We're not killing forests, we're saving them from disease. *We're not killing Vietnamese, we're protecting them from turning Communist, protecting them from themselves.* We're not killing forests, we're helping the local economy. *We're not killing indigenous peoples and stealing their land, we're developing natural resources to fuel the global economy.*

Deforestation has long been surrounded by a bodyguard of lies even more effective and widespread than beauty strips. Clearcuts become "temporary meadows" and "mimic natural disturbances." Clearcutting is called "even age management," or "mechanical fire suppression." Leaving a few trees in the middle of a clearcut is described as "selective cutting." Ancient trees are called "decadent," in the hopes that there will be less outcry over the loss of something already decaying than over the loss of something that was born long before our civilization and its war against nature. Old-growth forest is called a "biological desert" despite extensive scientific research showing that natural forests provide habitat for most of the world's threatened species.[1] Particular animal species are chosen as "indicator species" so that the entire forest ecosystem of interdependent species does not have to be considered as a synergistic whole.

Politicians, corporate journalists, and timber industry spokespeople often trot out lies and obfuscations to justify further deforestation. It is a measure of the insanity of our culture, the paucity of our discourse, and the corruption of corporate journalism that even the most absurdly transparent of these lies and obfuscations are rarely challenged, their inaccurate premises revealed. In order to inoculate readers against the most common of these lies, we've listed a few with their commonsense rebuttals.

Industry Statement: We have more trees today than we had seventy years ago.

First, note the use of the possessive *we have*. But trees don't belong to us any more than do water or air. They belong to themselves. Second, note that those in the industry begin their sample thirty or seventy years ago, after much of the forest had already been hammered by logging. That's a classic statistical trick: to narrow the window sufficiently to seem to make your point, then present this window as though it applies universally. For example, we've seen timber industry apologists write that increases in pronghorn antelope populations since early in the twentieth century mean "forest management helps wildlife."[2] This is just silly, not only because the plains-dwelling pronghorn antelope live nowhere near forests, meaning we may as well attribute their recovery to the success of the New York Yankees, but, more to the point, because as soon as Western culture arrived in the West, they commenced wiping out antelope, reducing their population from preconquest estimates of 10 to 15 million individuals in herds that rivaled those of the buffalo to a low of less than 27,000 in 1924. Since that time populations have recovered to about 700,000. If you begin your accounting in 1924, this seems a huge success—we've seen an increase of twenty-five-fold! If you broaden your analysis, however, you see the real effects of this extractive culture on these creatures: A reduction of population by 85 to 95 percent.

What's more, the spokespeople pull an even more disingenuous—and classic—statistical trick here, too, which is to conflate incomparable items as though they're identical. Wood fiber is not a natural tree, small trees are not big trees, and trees in plantations do not make a forest. It is not only absurd but obscene to conflate—as they're doing here—ten-inch seedlings to massive trees a thousand years old. Only 20 percent of the world's original forest survives as frontier forest—that is, relatively undisturbed forest in large enough tracts to be shaped by natural events and to support viable populations of native plants and wildlife. Seventy-six countries have already lost all of their frontier forest. Plantations don't count.[3]

In human terms, once a native forest is cut, it's gone forever. Many years ago, I said to Dick Manning—who wrote *The Last Stand,* about how he was fired as a corporate journalist for telling the public what was happening to the forests of Montana—how wonderful I thought it would be if we were to set aside more and more forests, and in five hundred years these might once again be old growth. He pointed out that first, many of the species have been at the very least regionally extirpated, so they might not come back, and more important, because some trees live five hundred or a thousand years, in five hundred years a forest will not have been even once through the nutrient cycle, with no trees having grown to old age, died, rotted, and become new trees. It takes thousands of years for a forest to become a fully functioning climax forest, with all the parts working together.

To even imply that a tree farm on a fifty-year rotation remotely resembles a living forest is either extraordinarily and willfully ignorant, or intentionally deceitful. Either way, those who make such statements aren't fit to make forestry decisions.

There are millions of acres of old-growth trees in the United States.

This is true, but misleading. There are 24.6 million acres of forest that are at least 150 years old—less than 5 percent of the total forest in the United States. There are 9.7 million acres of trees that are at least two hundred years old—less than 2 percent of the total forest area. For comparison, just one timber corporation, Weyerhaeuser, claims 5.7 million acres of timberland in the United States.[4] Plum Creek Timber claims 7.7 million acres in nineteen states.[5] International Paper controls 12 million acres in the United States.[6] Of course these timber company lands are no longer old growth. They are tree plantations, stands of second- or third-growth trees in varying conditions of health.

Our culture kills forests. The 20 percent of the world's original forest cover that survives as frontier forest generally does so because it is remote. Only 8 percent of the world's forests are under even nominal protection.[7] Worse, many of the surviving stands of ancient trees are fragmented by roads and are too small to function as habitat for wildlife. The sad fact is that even if they're spared from chainsaws, they're still not likely to remain in old-growth condition because of blow-downs and other sources of mortality to which small stands are susceptible.

The forests need to be cut to provide jobs.

The wood and paper industry and its markets are now global, with only a handful of companies left to compete. Over the past generation, employment has gone down as production has gone up. As companies continue to merge in order to reduce industry overcapacity and boost market share, they shed jobs. In the 1970s and 1980s, the number of paper mills in the United States decreased by 21 percent, but the average output per mill increased by 90 percent. Paper production in that period increased by 42 percent, while employment in the industry decreased by 6 percent. The amount of timber cut increased 55 percent, while the number of logging and milling jobs decreased by 10 percent, or 24,000 jobs. In just one decade (1987–1997), employment in pulp mills decreased by 2,900 jobs, and employment in paper mills decreased by 12,100 jobs.[8] Output per employee in the U.S. paper industry has increased fourfold in the last fifty years. The wave of consolidation in the pulp and paper industry that began in the late 1990s is expected to cost another 50,000 jobs.[9]

Any journalist with a shred of integrity would never have positioned debates in the 1990s over forest protection in the Pacific Northwest as "jobs versus owls," but instead perhaps "jobs versus automation, mergers, and downsizing." To frame the debate this way, however, would not serve the interests of those in power.

We need roads and logging to put out forest fires.

Fire is a natural phenomenon, necessary for healthy ecosystem functioning. Most natural forest fires are caused by lightning and they usually burn a very small area. Most large catastrophic fires are caused by industrial humans, often through the use of logging equipment or through slash burning following clearcutting. And catastrophic fires generally take place in areas that have been previously clearcut and replaced with unhealthy stands of crowded, small, even-aged trees. The Sierra Nevada Ecosystem Project Final Report to Congress put it succinctly: "Timber harvest, through its effects on forest structure, local microclimate, and fuel accumulation, has increased fire severity more than any other recent [sic] human activity."[10]

Forestry improves wildlife habitat, and in fact is good for forests.

As mentioned before, we need to ask what sorts of wildlife benefit from forestry—that is, from the conversion of forests to tree farms at best, and quite often to moonscapes. Creatures requiring forest interiors or large unfragmented habitat are consistently harmed. Anyone who does not understand—or who otherwise ignores—this distinction is qualified neither to make nor to influence decisions affecting the fate of forests.

Corporate journalists routinely parrot the industry line that industrial forestry is necessary for forest health. We've never seen any of them bother to ask the obvious next question: How then did forests possibly survive before the arrival of industrial foresters?

The industry is improving. Forestry operations are becoming more sensitive.

This is the classic line of exploiters everywhere, from perpetrators of domestic violence, to dictators, to heads of corporations. There may have been some problems in the past, but things have changed. We need to forget all that now, get on with our lives, and live for today.

There are some words in the domestic violence movement to describe those who believe these statements. They're called co-dependents, sometimes enablers, often victims, sometimes dead.

There are some words in the environmental movement, too, to describe those who believe the forest versions of these statements: they're called corporate journalists, timber industry hacks, and co-opted environmentalists.

Rates of deforestation continue to rise, populations of forest-dependent creatures continue to plummet. In many countries timber corporations often kill the people who oppose their depredations, and we're supposed to believe that the industry is improving? How stupid do those in power really believe we are? More to the point, if they themselves by some doubtful possibility actually believe these lies, how stupid must they be?

Further, and almost never remarked in the corporate press,

almost half of the cutting worldwide is illegal, not even obeying the lax environmental laws and timber royalty standards in place.

We need wood and paper products, so we need industrial forestry.

This is nonsense. The world is awash in wood and paper products. Much of what is manufactured is wasted. Much of what is manufactured is unnecessary—disposable cups and chopsticks, tissue, and packaging. Paper does not need to be made out of wood fiber, and rarely was until the last century.

The fundamental assumptions behind these wood and paper industry lies are that (1) the world needs industrial forestry to provide consumer products necessary for human survival, (2) clearcutting is similar to natural disturbances such as wind and fire, and (3) replanting a clearcut replaces a forest. The lies perpetrated by industry are perpetuated by the public's ignorance and unwillingness to understand forests. If people understand how the industry's fundamental assumptions are false, they are less likely to be fooled by the industry's other lies.

In response:

1. People and their immediate evolutionary predecessors lived for millions of years without consumer products. We do not need wood fiber paper packaging, disposable chopsticks and paper cups, cheap lumber, plywood forms for pouring concrete, and so on. The industrial model of forestry needs consumers, but people do not need industrial forestry. People do, however, need a livable planet.
2. Removing all the trees (and often all the other vegetation) from forest stands and watersheds is not natural. A very small percentage of the original forest cover was affected by fire or

other natural (including human) disturbances. Today, a small and rapidly shrinking percentage of the world's original forest cover remains. Thousands of forest species have already gone extinct due to industrial forestry—not evolution. The deforestation of the planet is not a natural process that has happened before. And once the destruction of natural forests is complete, they will not reproduce themselves in any comprehensible human time scale.

3. Plantations of fiber-producing trees grown with petrochemicals, and cut and recut in rapid succession, are not functioning forest ecosystems. Plantations of single tree species such as eucalyptus or pine are now grown in the tropics and cut every seven or eight years. Natural forests take thousands of years to evolve and stabilize, and require the symbiosis of thousands of animal, plant, algal, and fungal species.

A Rigged System

> The possibilities of power involved in such a concentration of land ownership, irrespective of the timber, hardly require discussion. The danger of abuse of that power, in the absence of restrictive regulation, is obvious.[1]
>
> *U.S. Bureau of Corporations, in its report on The Lumber Industry, 1913–14. (The U.S. Bureau of Corporations was disbanded soon after.)*

Nineteen ninety-five was the year I finally understood how the U.S. political system works, and at the same time realized how irredeemable are that system and the culture at large. That was also the year many indigenous friends said to me, "What took you so long to figure that out?" They'd had plenty of experience opposing this system—five hundred and some years of resistance to this culture and its environmental and cultural degradation—and had long since apprehended the truth in Red Cloud's words: "They made us many promises, more than I can remember. But they never kept but one. They promised to take our land and they took it."[2]

I was living in eastern Washington. I walked clearcuts there and in North Idaho that stretched for miles. No matter where I—or anyone—went, there they were. I saw streams scoured and essentially sterilized by "hundred-year" floods that came every few weeks in the spring. I saw migrating tundra swans dead, poisoned by lead from mine wastes flushed—sometimes to the tune of a million pounds per day—from the hills and into wetlands, rivers, and lakes by these floods. I saw politicians try to pretend nothing was happening as they scurried to protect the corporations that continued to cause the damage. (For years environmentalists at Forest Watch and the Inland Empire Public Lands Council in Spokane begged Tom Foley, their so-called representative and Speaker of the U.S. House of Representatives, to do something about the damage to the forests of his region. Finally, he deigned to say he'd take a look. The Public Lands Council arranged for a small airplane to take them all up. What happened next symbolizes much of what we're talking about in this book. Soon after the plane took off, Foley fell asleep. Two of the environmentalists kept waking him up and trying to get him to look at the clearcuts. He'd rouse himself long enough to yawn, rub his eyes, and glance outside before closing them again and returning to his dreams.) I saw corporate journalists stumble blindly over themselves, eyes clamped shut to see nothing wrong, suggesting that there was no need for concern over lead pollution, for example, even though some of the highest blood-lead levels ever recorded in human beings were from children in the area, because "there are no human bodies lining the Spokane River."[3] I saw populations of bull trout and Idaho cutthroat trout collapse. And through all of this the cutting continued.

There wasn't much we could do about cutting on lands controlled by corporations. Western culture—and this is an extraordinarily strange notion—values the "rights" of corporations—legal fictions, artificial constructs—above those of human, and especially nonhuman, life, and the landbase upon which all of this life depends. This value system meant we had only the most rudimentary tools available to us to slow the wholesale liquidation of forests on lands claimed by these legal fictions, lands to which, unsurprisingly, these corporations had for the most part gained title illegally from the public in the first place.[4]

But we did have some tools to slow the deforestation on public lands. The U.S. Forest Service and Bureau of Land Management sell trees to timber corporations at grossly subsidized prices. The prices routinely don't cover even administrative costs, much less reach market value for the timber, much much less repay the immeasurable costs of destroying forests, even when government timber-sale planners cook their books to a degree that would make Arthur Anderson proud. These administrators also push for environmentally destructive and publicly subsidized cattle grazing on public lands, environmentally destructive and publicly subsidized mining on public lands, environmentally destructive and publicly subsidized oil and gas exploitation on public lands, environmentally destructive and publicly subsidized ski resorts on public lands, and so on. You get the picture. You also get the shaft.

One of the differences between corporate and public lands is that public lands administrators have to at least give lip service to following stricter laws (except under certain conditions, as we'll get to later). They must maintain the façade of serving the public good. Part of that façade consists of writing documents called environmental assessments (EA) and environmental impact statements (EIS).

The ostensible purpose of environmental assessments is, obviously, to assess the environmental damage that will be caused by any "action" the Forest Service proposes to carry out. Not surprisingly, the Forest Service almost always determines that *every* action will have "no significant impact." This is true for such minor actions as installing pit toilets in campgrounds or hand removal of exotic plant species (each of which might have an entirely uncontroversial one-page assessment), and it's true for massive timber sales involving thousands of acres of clearcuts (which have several-hundred-page assessments inevitably coming to the conclusion of "no significant impact").

The much larger, more comprehensive environmental impact statements are written when the damage caused by the project is so extreme that not even the Forest Service can pretend there will be "no significant impact." The document then describes (read *understates*) the damage to be done.

By law, EAs and EISs are supposed to help the agency and the public make informed decisions about the "management" of the public's lands. As such, decision-makers within the agency are supposed to examine perhaps four or five alternative actions—which, in the case of a timber sale, may range from the "no action alternative" (no cutting) on one end, to clearcutting mile after square mile of forest and selling (below cost) tens of millions of board feet of timber on the other—and to choose among them the wisest course, based on the research that went into the document. But the system is rigged. The Forest Service and the Bureau of Land Management (as well as other federal agencies) often fire, threaten to fire, or otherwise make life difficult for biologists and botanists who find that activities such as logging, mining, off-road vehicle use, and gas and oil extraction harm forests. Just as routinely, they fire, threaten to fire, or make life difficult for cultural specialists who determine that any of these activities harm archaeological sites or sites sacred to Native Americans. The same happens to hydrologists who disclose damage to aquifers, streams, and rivers, toxicologists who disclose the poisoning of people and the landscape, and so on. Further, timber-sale planners and others are routinely moved to different forests every few years. This keeps honest planners from becoming attached to forests or communities, and guarantees dishonest ones a lack of accountability when their predictions of "no significant impact" are found to be false. The public does not even get the extremely hollow satisfaction of forcing the timber-sale planners to walk blasted streams. (Of course, if the planners were still there, they'd probably, like Tom Foley, simply close their eyes to the evidence anyway: After all, that's what they did through the planning process).

It will come as no surprise to anyone who has paid attention to American political processes—although I must admit at first it surprised me, naïve as I sometimes am—that EAs and EISs are not in fact documents designed to help people make informed decisions about *anything,* but instead attempts—often massive attempts—to justify decisions made long before, to satisfy back-room deals cut between politicians and their corporate backers. To be truthful, the Forest Service doesn't make a serious effort to even *pretend* EAs and EISs are actual decision-making documents: Out of the thousands of EAs and EISs from timber sales monitored and often opposed by groups with which I was associated, never once—*not even once*—did the Forest Service determine that the "no action alternative" was the best choice. The "preferred alternative" was *always* to deforest. Even a thousand chimpanzees typing on a thousand computers for a thousand years—the original line about this requires a million chimpanzees, but there aren't that many in the world anymore—would eventually come up with an EA that determined it was in the best interests of the forest not to cut. (Of course, being forest-dependent creatures, probably every EA the chimps wrote for proposed timber sales—were they foolish enough to develop a system in which timber sales were even considered—would point toward the "no action alternative.")

There are other ways the system is rigged, too, top to bottom. For example, general management of a particular national forest is at least ostensibly governed by what is called a Forest Plan. Often we would object to portions of the plans—primarily their emphasis on extractive industries—and would be told that our objections came at the wrong time: Instead we should appeal these points later on each specific EA or EIS. Then when we followed their instructions and waited to raise the points on EAs and EISs, we were told once again that our objections came at the wrong time: We should have appealed these points in the Forest Plan. Checkmate.

To appeal a timber-sale decision is a brain-busting process. The EAs and EISs are written in bureaucratese, which means it's almost impossible to figure out what the hell they're saying, if anything. The documents are intentionally deceptive—attempts to con an unsuspecting public into believing deforestation has "no significant impact." This makes the documents even more difficult to decipher. Terminology often changes. Each time we crack their old code for "clearcut," they come up with a new one, presumably in the hopes of sliding a bit more deforestation past us. Or maybe they enjoy destroying discourse as much as destroying forests.

We would read the documents, find the lies, determine the ways the Forest Service was violating such laws as the National Environmental Policy Act, the National Forest Management Act, the Endangered Species Act, the Clean Water Act, and so on. On one hand this was dead easy: the timber sales are glaringly illegal. On the other hand, the obfuscations made it tedious work to unravel. The work was made all the worse by the knowledge that those tying these words into rhetorical knots were pulling thirty, forty, or fifty grand per year to confabulate the lies while most of us doing the unraveling weren't paid anything at all (a couple people on our side *were* getting paid, a whopping $16,000 per year). We were doing it for love, and because it was right. It is fun and fine and wonderful to do things for love, but it galled me no end that those destroying the natural world were getting paid for it while we had to try to stop their damage for free. If we had to describe the pathology of our culture in a nutshell, that might be it: our economic system rewards destructive behavior.

There is one night I'll never forget. It was winter. About 1:30 in the morning. Clear. Bright stars punctured the sky. The cold so sharp it pierced my cheekbones. I'd been tearing apart an EA since 6:00 that evening, and I knew I had to quit when I saw a chart on page 175 with a caption stating it was identical to one on page 43, "reproduced here only for readers' ease." But the charts were different. They were being used to make different points (the chart on page 43 was supposed to show how few cutthroat trout were in a stream, and that on page 175 was supposed to show the opposite). The authors had made up different charts—and different underlying data—to make their different points. I threw the EA across the room, put on my coat, and stalked into the cold.

I walked long through the night air, trying to unlock my brain, get it to stop spinning in fast tight circles that were making me ill and tired. I wanted to quit not just for the night but forever. I did not want to submit myself to this abuse.

But I knew I couldn't do that. That's what they wanted, to wear us down with their lies. And I wasn't going to let that happen.

We often received late-night calls from dissident Forest Service personnel. Someone might call from what sounded like a pay phone, not give a name, and say, "Look very carefully at page 57. Whatever you do, don't forget to notice the lack of analysis of the effects of this timber sale on both goshawks and Thompson's big-eared bats." Then the person would hang up.

We'd write the appeals, and send them, oddly enough, back to the same people who signed them in the first place. They would of course deny the appeal, and we would appeal their denial to their supervisor, who would of course then deny our appeal. We'd play this game all the way up the line, until someone finally granted the appeal or we took them to court. Sometimes we couldn't take them to court because we didn't have the money to throw at an attorney, but fortunately there were a fair number of attorneys willing to help us fight these battles, often for free or on the cheap.

But as is always the case when attempting to stop our culture from destroying some part of wild nature, all losses are permanent, all victories temporary. Winning a timber-sale appeal doesn't mean stopping a timber sale. It doesn't mean protecting a piece of ground. It means protecting a piece of ground for the year or two it takes the Forest Service to write up another EA, this time trying harder to bamboozle us. The score today is that less than 5 percent of the ancient forest in the United States remains.[5]

Despite the fact that the whole system is rigged in favor of deforestation, a bunch of us were able to use the system's own rigged rules to stop, for a while, most of the illegal logging on the national forests of our region. Because nearly all commercial logging on public lands violates environmental protection laws, this means we stopped nearly all commercial logging on these forests. Activists across the country used similar tactics with similar success to try to enforce the law and protect the forests.

The response by our local Forest Service was to hire scores of new employees. Were these new hires biologists, botanists, hydrologists, and anthropologists, brought on in an attempt to better understand forests? No. They hired one timber-sale planner, and all the rest were technical writers directed to produce slicker documents.

The response nationally was rather more severe, and is what helped me learn how unwaveringly committed to the destruction of the planet our culture is. The timber industry, politicians, and the corporate media launched a massive propaganda campaign. This in itself is nothing new: It's what they all *do*. But they took the momentum we had gained in our descriptions of devastated forests and inverted it to declare: *The forests are suffering a major health crisis, so we need to move quickly to cut them down.*

Stop. Reread their declaration. Read it again. I've read that or similar lines too many times, and they still make no sense.

But that's one of the advantages of wielding the sort of totalitarian power held by the government/corporate interlock. While it's certainly more convenient for them to carry out their "preferred alternative" without too much public resistance, public assent to their goals is ultimately incidental. Any time the public finds a way to meaningfully participate in decisions concerning its own landbase—as we did with the appeals process—they simply change the rules. Almost any excuse will serve to allow them to sever public participation in the process. As we've seen time and again, on issue after issue, when the corporate press publishes absurdities often enough, the absurdities begin to seem palatable to some, confusing to others, and discouraging to still others. As long as it paralyzes the public, the corporations win. Ninety-five percent of the old growth is gone, and they're getting away with cutting the rest.

So in 1995 Congress passed and President Clinton signed what became known as the Salvage Rider, which stated that the quickly worsening health of the forests demanded immediate action. Therefore, any timber sale that the Forest Service or Bureau of Land Management (BLM) declared necessary to improve "forest health" would be exempt from all environmental laws. The Salvage Rider contained something called "sufficiency language," a magical phrase meaning no appeals or other legal challenges are allowed. Public participation is explicitly prohibited. Of course.

Can you guess what happened? The Forest Service and BLM predictably declared nearly every timber sale to be necessary for forest health. It was a chainsaw massacre. Ancient forests fell everywhere. In my corner of the world, every one of the thousands upon thousands of acres I had worked to save—every goddamned acre, every fucking acre, every beautiful, vibrant, stunning, gracious, wise, living acre of ancient forest—was clearcut over the next two years. I did not have the courage to return to many of those places. I could not have borne to see them destroyed. Others I did go to see, and walked the moonscapes that until recently had been living, vibrant forests. It is not an experience I look forward to repeating, though of course it is an experience shared by all of us who love wild places and who are facing down the deforesters.

This is how our political system works. Choose your own equivalent example; they are myriad. This is why the system must go.

In one sense the whole salvage hoax was unbearably stupid. It doesn't take a genius to figure out that if any of the logging had been truly necessary to improve the health of the forest—health that was already being damaged by, you guessed it, logging—there would have been absolutely no reason to exempt it from environmental laws. In another sense the hoax was pretty clever, in that it took a genuine fear and turned the cause on its head. That's standard practice in propaganda. You don't try to make something out of nothing, because then your lies have no energy. Instead, you rechannel existing energy—fear, desire, anger, whatever—toward your own ends, your "preferred alternative." So, Nazis took the very real anger of a defeated and humiliated people, and the fear of these same people facing an economy in chaos, and rechanneled that energy toward their own deathly ends. Those in the U.S. military-industrial complex twisted people's real desire for security into a military machine designed to achieve, according to the military's own Joint Vision 2020 statement of purpose, "full-spectrum domination." Those in the timber-political complex turned the energy from our efforts to halt deforestation back against us, or rather against the forests, with the "forest health" scam and subsequent Salvage Rider.

Now they're doing the same thing with forest fires, inflaming public fears as they did with forest health, using partial truths and often outright lies to turn some very real concerns to their own destructive ends. Fires are a normal part of forest ecology, especially for many of the forests of the arid west. In fact many species are fire-*dependent*. Lodgepole pines, for example, are shade-intolerant and need openings in the forest canopy such as those caused by fire to send up the new generation. To this end, lodgepole pinecones are tightly closed by resin that only fire can melt, and seeds can only germinate on exposed soil, where fire has removed leaf litter. Three-toed woodpeckers, also called black-backed woodpeckers, are colored to be camouflaged against the charred background of burned trunks. They exist only precariously where there is no fire, but arrive in droves to burned areas to eat bugs feeding inside blackened trunks.

Fires are a forest's way of renewing itself. In arid forests, fire, not bacteria or fungi, is the main agent for breaking down nutrients. Without fire, dead litter in these forests does not decay or rot, but simply piles up on the forest floor.[6] Fires also redistribute these nutrients across space—in the form of windborn ash—much as salmon carry nutrients from ocean to forest. Fires mix things up: They're a tremendously creative force.

And for the most part fires aren't all that dangerous. I know we've been raised on stories of sad Smokey the Bear clinging to a tree, his mother obviously murdered by the raging fires. He was saved (and imprisoned) by kind humans in green polyester pants. But who told us these stories? Those kind humans in green polyester pants themselves, the members of the Forest Service.

Would they lie to us?

Well, yes.

Most natural fires are pretty small—far less than 100 acres; even including larger fires the average is only about 240 acres—and they don't burn quickly or all that hot.[7] They don't jump to the tops of big trees, but kill only their smaller cousins beneath. And there are a lot of these small fires: The Blue Mountains of Oregon got their name from the smoky haze of so many small wildfires. In their natural cycle most western forests burned every three to twenty years, with longer cycles in more moist forests. This means big trees in dry forests would experience perhaps fifty or a hundred fires in their lives.[8]

These small fires aren't terribly dangerous. Because the fires burn in patches, animals easily move to protected swales until the fire passes (or they climb trees and wait for mother to return, and hope she gets there before the bastards in the green polyester pants). The front usually advances at only a couple of miles per hour, meaning large mammals can easily amble in front of the flames, and birds can fly away. (Isn't it cool, by the way, how most birds raise young early in the year, which means fledglings are ready to fly before fire season?) Even small creatures are fairly safe from fires: Mammals head into their burrows, and insects just dig themselves into the soil, where a few inches below the surface the temperature remains remarkably constant.[9]

The nature and danger of forest fires changed with the arrival of extractive forestry. Peshtigo, Wisconsin, October 8, 1871: only twenty years earlier the area had been part of a 200,000-square-mile unbroken native forest that covered much of Wisconsin, Michigan, and Minnesota. But the trees were cut, for lumber, for railroad ties, to clear land for agriculture. Fires escaped from logging slash piles and exploded into the logged-over forests, burning 1.25 million acres of pine trees, and killing 1,500 people. Hinkley, Minnesota, 1894. Metz, Michigan, 1908. Cloquet, Minnesota, 1918. These huge fires raged in the aftermath and as a result of extractive forestry.[10]

As extractive forestry moved west, so did catastrophic fires. The Yacoult Fire of 1902 (actually a series of 110 fires), started by logger and settler fires, burned a million acres in Washington and Oregon and killed thirty people. Then came the Wallace Fire of 1910 in Idaho, sometimes called The Big Blowup. As usual, logging created conditions ripe for catastrophic fire—lots of slash piles, lots of trees killed by logging, lots of weak "dog-hair" trees coming up in dense even-age stands. By July of that year 3,000 fires—many started in slash piles—were burning in the forests of North Idaho. On August 20, "all hell broke loose," according to the District Forester: hot hurricane-force winds blew up from the southwest. They were strong enough to blow riders from their saddles, and strong enough to bring together the small fires into a conflagration that took out 3 million acres of white pine. Headlines from the region: "Wallace Fire Loss $1,000,000: 50 Dead—180 Missing in St. Joe Zone"; "Five Known Dead Near Newport"; "Terror-Stricken, 2000 Refugees Dash Through

Flames to Safety"; "In Forest Fires 142 Dead, 185 Missing, Property Loss is $20,000,000"; "Fire Victims Number 185."[11]

In response to the destruction caused by these logging-induced forest fires, the federal government decided to head right to the root of the problem and halt all industrial forestry, right? Well, no, not exactly. Instead administrators decided to eradicate not the disease but the symptom, and assumed what was called a "10 A.M. fire policy"—every fire must be out by ten o'clock the next morning. Inmate Smokey was conscripted into serving the propaganda effort on the part of the Forest Service to sell this policy to the people of the United States.

The net effect of this policy has been a further weakening of already stressed forests, as well as a dangerous buildup of those even-age stands so beloved of both foresters and fires. In other words, industrial forestry has combined with a misguided fire-suppression policy to create conditions ripe for disaster.

The federal government has, unsurprisingly, used this fear of conflagration to promote deforestation by greatly increasing logging and by suspending environmental laws and public participation on all timber sales determined to be necessary to "reduce fuel load." The Forest Service and the Bureau of Land Management have already put their old rubber stamps bearing "Necessary for Forest Health" into storage and replaced them with stamps bearing "Necessary to Reduce Fuel Load."

Expect another massacre. It doesn't really matter to the outcome that time after time the trees that are cut are the big old commercially valuable trees, not the dog-hair trees more prone to fire. A 1999 Government Accounting Office report stated that Forest Service managers "tend to (1) focus on areas with high-value commercial timber rather than on areas with high fire hazards or (2) include more large, commercially valuable trees in a timber sale than are necessary to reduce the accumulated fuels." Nor does it matter that a September 2000 report by the Department of the

Interior and the Department of Agriculture stated, "The removal of large, merchantable trees from forests does not reduce fire risk and may, in fact, increase such risk." Nor does it matter that Forest Service fire specialist Denny Truesdale says, "The majority of the material that we need to take out is not commercial timber. It is up to three and four inches in diameter. We can't sell it."[12]

Science doesn't matter. Logic doesn't matter. Public participation and democracy don't matter. Justice doesn't matter. Forests don't matter. Life doesn't matter.

Commercial logging removes large, fire-resistant trees and leaves behind flammable needles, limbs, and brush. What's more, removing the overstory reduces shade, drying and heating the materials below. Tree plantations are far more vulnerable to fire than natural forests, and there is a direct correlation between roads and fires. Add to this the fact that the overwhelming majority of forest fires—88 percent—are caused by humans, and that up to half of these are arson. There have already been many cases of people lighting fires *specifically* so they can benefit financially, whether through gaining employment as firefighters or through giving the Forest Service an excuse to offer up the dead trees as a timber sale, quite possibly to the arsonists themselves. The Forest Service's own fire laboratory found that the main factors determining whether buildings ignite are the materials used in the home and the amount of underbrush within two hundred feet, *not* the merchantable timber within two hundred miles. But to those in power who are deforesting the planet, it positively does not matter that study after study after study has shown that logging leads to catastrophic fires.

We'll tell you what does matter. Logging, under the guise of forest health, under the guise of reducing fuel loads, under any guise those in power claim, under no guise at all but just because those in power make the rules, serves the interests of the big timber corporations (which, not coincidentally, recently exercised "their first amendment right" to free speech to the tune of more than 3 million dollars in payola—sorry, campaign contributions—to the presidential campaign of George W. Bush).

This is how the U.S. political system works. This is why the system must go.

Corruption

> The battle we have fought, and are still fighting, for the forests is a part of the eternal conflict between right and wrong. . . . So we must count on watching and striving for these trees, and should always be glad to find anything so surely good and noble to strive for.
>
> *John Muir, November 23, 1895*

Part of the reason government and industry work so closely together is that they're parts of the same machine, working for the same ultimate purposes. One primary purpose is to maintain production—to convert forests into chopsticks, two-by-fours, and newspapers. Another way to say this is that a primary purpose is to convert the living to the dead.

Another reason government so often supports industry is that a revolving door exists between the two. Politicians and bureaucrats were often pillars of industry before they went into politics or "public service," and after politicians get booted out of office, where do they go? Back into the private sector. This revolving door provides sophisticated incentives for future and ongoing career opportunities, salaries, bonuses, and other benefits. It's no wonder those in industry and government can believe and say, "What is good for the country is good for General Motors, and what's good for General Motors is good for the country." Having already defined America not as its land or citizens but as the government, they're simply saying that what's good for themselves is good for themselves.

Thus, soon after Lee Thomas left his job as head of the United States Environmental Protection Agency (EPA), he joined Georgia-

Pacific, one of the companies he had pretended to oversee.[1] William Ruckleshaus, who also headed the EPA, went on to sit on the board of Weyerhaeuser, Browning-Ferris Industries, Cummins Engine, Coinstar, Monsanto, Nordstrom, Solutia, and Gargoyles.[2] Sometimes the door doesn't even need to revolve: Booth Gardner, former Washington State governor and U.S. ambassador to GATT, is a multimillionaire heir to the Weyerhaeuser fortune. Gardner was declared exempt from a 1972 Washington State voter initiative requiring public officials to disclose their financial assets; the Washington State Public Disclosure Commissioners—appointed, conveniently enough, by the governor—renewed his exemption every year.[3] Charles Simon, a chief researcher for the industry front group National Council of the Paper Industry for Air and Stream Improvement, was later a consultant to the U.S. government in its investigation of the paper industry's violations of pollution laws.[4] U.S. senators such as James McClure and Slade Gorton—two of the most genocidal and ecocidal American politicians since Andrew Jackson—left "public service" to join law and lobbying firms that cater to timber and mining corporations receiving public tax monies and public lands resources.

The list is long. Who better to oversee the Forest Service than the attorney who defended Louisiana-Pacific from charges of monopolistic practices detrimental to the people and forests of the United States? Ronald Reagan appointed such a person—John Crowell—Chief of the Forest Service. Crowell immediately set a goal of doubling timber production from the national forests by the year 2002. That didn't happen, in part because there weren't that many trees left to cut, even if the market could have borne all that wood. But the cut did increase, until by 1988 the United States became a net exporter of wood products for the first time, and Americans were subsidizing the Forest Service's destruction of public forests with billions of tax dollars.

Mark Rey's career is a more recent example of the revolving

door. In the mid-1970s he worked for the Bureau of Land Management, and then in the late 1970s and 1980s he worked for the American Paper Institute, the National Forest Products Association, and the American Forest Resource Alliance, a "Wise Use" industry front group. By the early 1990s he was a vice president for the American Forest and Paper Association.

Then Rey returned to the government, as a staff member with the U.S. Senate Committee on Energy and Natural Resources. Rey designed the infamous Salvage Logging Rider of 1995 (written directly after he "left" the industry for "public service") and worked on the Herger/Feinstein Quincy Library Act of 1998 and the Secure Rural Schools and Community Self-Determination Act of 2000 (the former got out the cut under the guise of "forest health," while the latter got out the cut under the guise of providing money for schools and economic stability for communities). He also drafted Senator Larry Craig's rewrite of the National Forest Management Act to eliminate citizen oversight committees and other environmental protection measures. The interesting thing about this bill is that many of its recommendations replicated word-for-word statements made by the man who succeeded Rey at the American Forest and Paper Association.[5]

At a speech given at University of California at Berkeley in October 2000, Rey stated: "Our public lands are now under the protection of sweeping laws, like the Endangered Species Act, enforced by powerful federal agencies. There is no emergency that warrants this unilateral exercise of executive authority."[6]

Rey believes—or at least states—that clearcutting is "compatible with rain forest ecology," "relatively comparable" to windstorms, and benefits wildlife by clearing out dense sections of forest for animals. He stated that the 1991 Fish and Wildlife Service proposal to protect spotted owl habitat was an "insane proposal [to] place the interest of owls above the interest of thousands of logging families and communities." (Of course he did not suggest that automation is an "insane proposal to place the interest of corporations above the interests of thousands of logging families and communities," although automation costs far more jobs than environmental protection.) He also suggested limiting the Forest Service's budget to "custodial management."[7]

Part of Rey's reward for such antienvironment and antigovernment statements was to be named by President George W. Bush as Undersecretary of Agriculture for Natural Resources and Environment, overseeing the Forest Service and the national forests. He is well-placed and well-connected to continue his influence over forest management in the United States. "Rey's strength in shaping timber issues comes from his well-nurtured contacts within the federal agencies, with the media, with labor unions, and within the timber industry."[8] The Democrats went along with Rey's appointment, too; it was unanimously confirmed by the Senate Committee on Agriculture, Nutrition, and Forestry, and then by the full Senate.[9]

The door revolves not only between government and industry, but also between industry and big environmental corporations. Jay Hair left his cushy job as President and CEO of the National Wildlife Federation to become a PR flack for Plum Creek Timber Company, a company so outrageously environmentally destructive that even a Republican congressperson called it "the Darth Vader of the timber industry." That the National Wildlife Federation even has a CEO is a sign of how corporate it has become. Linda Coady, formerly a senior executive at Weyerhaeuser, became vice president of the World Wildlife Fund's Pacific regional office. Corporations also directly fund these large organizations.[10] It's no wonder large "environmental" corporations such as Audubon, Sierra Club (which threatened to expel members who spoke out against President Bush's invasion of Iraq), Environmental Defense Fund, World Wildlife Fund, and Natural Resources Defense Council are more interested in raising money than in raising discomfort among the economically powerful. It's no wonder they're more interested in defending bottom lines than forests. It's no wonder they consistently undercut the efforts of true grassroots organizations working to protect their homes: What's good for GM is good for America is good for the corporate environmental organizations.

Republicans do not, of course, have a monopoly on influence; in our system, corruption is a bipartisan effort. We spoke earlier of the Forest Service Timber Theft Investigations Branch (TTIB), but it might be appropriate to now explore it a bit more. If you recall, the TTIB (until it was disbanded in 1995) showed that there was "collusion by top Forest Service officials (1) in after-the-fact authorization to illegally cut timber; (2) after-the-fact 'new math' to excuse industry failure to pay for hundreds of thousands of dollars of timber previously cut; and (3) de facto warnings to company targets of an ongoing probe, and wide distribution of confidential case information."[11]

TTIB's investigations "revealed collusion between certain district managers and the timber industry that has resulted in the robbing of millions of board feet of timber from our national forests,"[12] and the Forest Service has admitted that up to 10 percent of all trees cut from the national forests are stolen, costing taxpayers up to $100 million a year. Theft of public trees is so rampant and so encouraged—or at the very least not discouraged—that some have taken to referring to the Forest Service as the "unindicted co-conspirator" behind every timber fraud case. We suspect 10 percent poaching is an underestimate.[13]

The TTIB interviewed witnesses who claimed that up to a third of the trees on some timber sales in the Tongass National Forest were cut illegally. "Millions of board feet of top quality lumber were falsely graded as worthless cull; loads of high value tree species were disguised by placing lower value trees on top; dozens of log rafts, each worth a million dollars, were diverted before reaching the scaling yard; records were missing or incomplete, in some cases skipping counts for up to half the logs in a sale."

Further, logs were illegally exported from the Tongass. Rafts of logs were regularly routed to a port under American Indian jurisdiction where there was no functional Forest Service presence. At night these logs were secretly shipped to foreign markets in Japan, Korea, and Taiwan. The thieves were so brazen, the Forest Service oversight so slight, that unmilled logs were often shipped—illegally—out of the main port, Thorne Bay, in broad daylight.[14]

But as mentioned before, TTIB was suddenly disbanded in 1995, just as investigators were closing in on Weyerhaeuser and other corporations. The important thing to mention here is that the 1995 abolition was delayed until just after Dan Glickman completed his confirmation hearings for Agriculture Secretary: another case of politicians delaying the delivery of bad news until after their re-election, and for the same purpose. This surprise announcement was then publicly explained as the chance to expand a regional investigative unit into a truly national commitment. But that's another beauty strip: Precisely zero agents have been trained in corporate timber theft, and the average value of prosecutions has dropped to $1,500, the equivalent of some stolen firewood and Christmas trees.[15]

Money is the most important thing in our culture (with the possible exception of the power this money represents, and that is necessary to the acquisition of ever more money). The interlocking staffs of government, industry, and big environmental organizations are united by personal profit, and the revolving doors between are greased with cash.

The American Forest & Paper Association (AF&PA) consists of wood and paper corporations and trade associations promoting "a policy and economic climate to remain competitive worldwide."[16]

AF&PA and its member corporations gave federal politicians more than $8 million in legal political contributions from 1991 through June 1997. Top contributors included International Paper and Georgia-Pacific, each of which gave $1 million, and Stone Container and Westvaco, each of which gave $600,000. In return, AF&PA members received more than $100 million in discounts on national forest timber during that same period. Major beneficiaries of the Forest Service "road credit" (tax subsidy) program included Sierra Pacific (receiving $20 million), Boise Cascade ($18.9 million), Willamette Industries ($8.8 million), Weyerhaeuser ($7.5 million), Stone Container ($5.3 million), Plum Creek ($4.6 million), and Potlatch ($4.2 million).[17] Clearly, the highest rate of return any corporation can attain is through the purchase of politicians.

The cozy relationships and corrupt transactions that drive U.S. forest policy and deforestation also occur in and between countries, and are the foundation of the international wood and paper trade. Industrialism (whether socialist or capitalist) requires cheap and obedient labor, a constant input of raw materials, and ever expanding markets of indiscriminant consumers. Being parasitic, the system requires continued subsidy by nature and the capture of ever more customers. "Globalization," then, is the spread of this parasitic, monetized, commodity-driven, inequitable, monocultural socioeconomic system from the center of empire to its periphery.

One global deforestation story among many has to do with Japan and the forests of Southeast Asia. Some of the primary drivers of the deforestation of Southeast Asia over the past two generations have been Japanese trading companies *(soga shosha)*. The largest of the trading companies are Mitsubishi, Mitsui, Itochu, Sumitomo, Marubeni, and Nissho Iwai, but there are many more. The *soga shosha* are huge networks (Itochu has eight hundred subsidiaries and affiliates) of suppliers, providers, and customers that collectively account for 44 percent of Japan's imports, 30 percent of its exports, and 25 percent of the Japanese gross domestic product. The trading companies provide financing, market information, technology, and expertise, and arrange production, shipping, and supply contracts. In the case of wood and paper, a trading company might finance and facilitate contracts between loggers, shippers, exporters, plywood manufacturers, wholesalers and retailers, and construction company customers. Japanese trading companies are unlike traditional Western corporations in that they do not seek the highest profits. Instead, their profit is based on huge volumes. They accept low fees as their revenues, and stay in business by providing credit and equipment for cheap logs, controlling trade chains to stimulate demand, and using transfer pricing and other schemes to evade taxes. The result is a massive supply of low-priced pulp, paper, lumber, and plywood. The trading companies sweep across Southeast Asia, decimating tropical forests as they go: To the Philippines in the 1950s and 1960s, Malaysia and Indonesia in the 1970s and 1980s, and then on into Papua New Guinea and the Solomon Islands.[18]

The Japanese trading companies' assault on tropical forests has been facilitated by local versions of the patron-client relationships mentioned in the discussion of the U.S. timber industry. Patrons such as Indonesian President Suharto provided security, state resources, licenses, and timber concessions to political, bureaucratic, military, and business clients, who in return provided Suharto and his allies with political support, financial backing, legitimacy, and stability. One client of Suharto's, Bob Hasan, controlled the Indonesian plywood trade. Suharto's Chief Ministers Taib Mahmud and Rahman Yaakub controlled almost a million acres, a third of Sarawak's forests. Other clients enriched and protected by Suharto included the owners of the world's largest pulp and paper mills, which have destroyed millions of acres of Indonesian rainforest to supply Japanese and American consumers with cheap photocopy paper.

Many of the logging operations and plywood and paper mills that have destroyed the tropical forests of Asia were financed with loans from Japanese banks and government agencies. The borrowing countries could repay loans only with revenues from the logging, but the cut-rate prices set by the companies as well as widespread corruption that allowed loggers and buyers to escape timber taxes and fees meant more and more forest had to be cut to keep up with payments. The countries incurred more debt to expand operations. Japan was the largest source of "foreign aid" money in the world, and dwarfed all the regional development banks combined. Japan's "environmental" aid turned out to be yet more loans, which were used to inventory the remaining forests for logging, and to "reforest" with pine and eucalyptus plantations for further fiber production.

Because patron/client relationships are between parties of unequal power, they are by definition unstable. Paranoia, suspicion, possible exposure of corruption, and falling out of favor always hang in the background. Of course the luxurious lifestyle of the elite and the poverty of the people fuel resentment. So inevitably, Suharto and his cronies fell. Unfortunately, his fall did not halt the dispossession and deforestation. By this time, Indonesia had racked up an immense national debt to banks and international financial institutions such as the World Bank and the Asian Development Bank. Much of this money had gone into building infrastructure necessary for transnational corporations to cut forests, to mine minerals—in short to remove Indonesia's resources. Indonesia was then left—and we see this pattern the world over—with a devastated landscape as well as a huge debt: Both trees and cash flew out the door. Those at the center of empire use debt to force more deforestation: *If you don't have handy cash, we'll be kind enough to take trees. And by the way, why are you wasting so much money on schools, hospitals, and environmental remediation?* It's a handy scam, the impolite term for which is loan sharking. Suharto the dictator fell, but the dictatorship of the international financial institutions continues to take resources from the poor.

Similar structures and processes occur in Africa, where the former colonial powers such as England, France, and Belgium continue to siphon off the forests of their former colonies—and half or more of the tropical timber they import is from illegally logged sources.[19]

The same thing happens today in Brazil. Brazil was, as you know, conquered by Europeans. The forests were depopulated by murder, disease, and forced labor, and those who remained were turned into peasants. They were dispossessed. Today the dispossession continues. Less than 1 percent of landowners control 43 percent of the land, and 6 percent hold 80 percent of the farmland. Sixty-eight million acres of that farmland is idle, held for speculative purposes (the Madeireira Manasa company, for example, employs only sixty-eight people on its million and a half acres). Transnational corporations hold 14.5 million acres of Brazilian land. This in a country where 86 million people (two-thirds of the population) are undernourished, and *this* in a country that is the second-largest exporter of agricultural commodities (soybeans for European cattle, coffee for Americans, sugar for the coffee).

What does all of this have to do with deforestation? The largest twenty landowners hold 8 million acres. They are senators, ministers, and army chiefs. The Brazilian government, controlled by these bigwigs, orders the massive resettlement of the landless poor onto newly logged Amazon rainforest, despite the fact that less than 10 percent of the soils in the Amazon can sustain annual food crops. Here's how it works: Those in power order the forest cleared to obtain timber, oil, and minerals. Then the restless land squatters are released from overcrowded cities. Soils are destroyed by farmers desperate to make a meager living. The frontier is pushed back farther. These resettlement schemes are financed by the World Bank and International Monetary Fund, as well as by North American and European timber, oil, and mining companies. Those who run these companies and who make these policies see forests and those who live there as natural resources, raw materials, commodities—things that can be killed to turn a buck.

A reform program that promised to redistribute land ownership in Brazil in the 1960s was scuttled by a U.S.-backed coup. A generation later, during the 1985 transition to civilian government, rural violence against the poor increased yet again, perhaps as a warning to those who saw a chance for commonwealth as well as freedom. The largest landowner in Brazil was the Minister of Agriculture. Ministers, state governors, mayors, and judges helped raise money for terror killings. When Catholic priests began preaching liberation theology, some of the landlords converted to Protestantism, and some of the priests were murdered. The Vatican condemned liberation theology, and imposed silence on its adherents. The Trans-Amazonian and Rondonian Highways (funded by the World Bank and Inter-American Development Bank) were supposed to help resettle more than a million Brazilians, but they destroyed millions of acres of forest, and the resettled were later displaced for mechanized soybean and other cash crops. Thousands starve, while thousands more have been murdered for squatting on idle land or organizing labor unions.

We can tell similar stories of countries, people, and forests the world over. Take Guatemala—the transnational corporations certainly have. As Marcus Colchester and Larry Lohmann put it in their book *The Struggle for Land and the Fate of the Forests,* "The violent subjugation of the [Guatemalan] Indians was . . .

integral to the imposition of export-orientated economy that the metropolitan centre required. The extent to which the Indians resisted, by all possible means, this enforced assimilation into the market economy has not been sufficiently appreciated. One indication of its depth may be gauged from the fact that there has been an average of one Indian rebellion every sixteen years since the conquest of Guatemala in 1524 to the present day."[20]

These rebellions have been met, and continue to be met, with force. Every attempt by the people of the forests of Guatemala to maintain control of their lives and landbases, or at the very least to hold on to the value of their own labor, is met with goons and gunboats and coups, and perhaps even more deadly, bankers in suits wielding the full force of the law. And there has been no shortage of mechanisms in Guatemala (and elsewhere) to move the indigenous off their land and into the labor pool. These include the (often-forced) privatization of communal lands; tax schemes that force people out of subsistence and barter economies and into the cash economy (which means into the wage economy: how do you pay taxes if you grow food only for your own family, plus enough to barter with neighbors?); vagrancy laws that force people to pay rent, which is a way to force them to get jobs; and racial requirements for land ownership. All of these have guaranteed the rich not only access to resources but access also to landless peasants for cheap labor. Indeed, Guatemalan Indians have been routinely sold along with the haciendas.

The government of Guatemala—and this is really true for all third world governments—has been forced into an unworkable situation. The country's experience with United Fruit (UF), an American transnational corporation, shows how people and governments of the colonies are kept in place. United Fruit arrived in Guatemala in 1899. By 1930, UF was the largest landowner and largest employer. In 1931 the government, at the urging of large corporations, decided to break traditional communal lands into private plots, including Indian lands UF wanted to exploit. In 1954, when United Fruit faced the nationalization of 387,000 acres of land—the government was, scandalously enough, going to pay UF the precise value the company claimed the land was worth on its tax rolls—the Guatemalan government was overthrown in a U.S.-backed coup, leading to thirty years of bloody dictatorship, and the murder of at least a half a million Guatemalan Indians.

In 1960 three-quarters of Guatemala was covered by forest. In the next decade, as a result of oil money and international loans, cattle ranching increased (even as domestic beef consumption fell by 50 percent: the average American housecat eats more beef than does the average Guatemalan). Agribusiness, oil exploration, nickel mines, and hydroelectric projects pushed back the frontier. Indian massacres were followed by colonization schemes transferring even more land to the middle and upper classes and to the military leaders who oversaw the massacres. Two percent of the farms now enclose two-thirds of the land in the country. Much of the land is idle, guaranteeing cheap labor. Eighty-eight percent of the farms are too small even to support a family. There are 300,000 landless laborers without permanent jobs.

By 1990, about a quarter of the country was forested, and perhaps 2 percent of the original frontier forest survived. Some specifics: The Peten forest once covered the northwestern third of Guatemala. Between 1964 and 1984, a third of the Peten was deforested, the population increased from 27,000 to 200,000, and natural resources came to be controlled by the military.[21]

Eighty-six percent of Guatemalans live below the official poverty line. Half of the Guatemalan children show retarded growth. Infant mortality runs eighty per thousand births. Subsistence standards are now lower than during the colonial era 350 years ago.

What is the response by those in power? Well, just as increased logging, we're told time and again, will cure logging-induced forest health and fire problems, so, too, according to the World Bank and the Inter-American Development Bank, will debt, deforestation, and exporting resources help Guatemalans. Money is pumped into agricultural factories—they're not farms, really—that produce broccoli, peas, melons, berries, and flowers for export. These crops require lots of water and petrochemicals. And they require lots of soils: Typically, an acre that is denuded of trees and heavily cultivated can lose five to thirty-five tons of topsoil per year. The people starve. The forests and people die.[22]

Or we can talk about Cambodia. Thirty years of war, ten years of United Nations aid embargo, the horrors of the Khmer Rouge—all took their toll on forests. Every side in the wars used illegal logging to fund the fighting, with much of the timber being smuggled out through Thailand and Vietnam. No matter who won battles, forests lost the wars. The country's forest cover went from 70 to 30 percent, and perhaps 10 percent of Cambodia's original frontier forest remains.[23] In 1995 the Cambodian government conceded the remaining forests to thirty companies. That did not even help the people of Cambodia financially: In 1997, when Cambodia's annual budget was $419 million, an estimated $185 million worth of timber was cut, but only $12 million went into the treasury. Logging increased in 1998, but Cambodia received only $5 million. Government and military leaders facilitate the smuggling. And where does this wood end up? Much of it is made into European garden furniture, some even marked "environmentally friendly wood products" made from "sustainable harvested plantations." Exporters and brand names include ScanCom and Tropic Dane (Denmark), Eurofar and Unisource (Holland), Comi and De Bejarry (France), Cattie (Belgium and France), Robert Dyas Ltd and Country Gardens Centre (UK), Beechrow (Australia), and Sloat Garden Centre (USA). The Worldwide Fund for Nature's 95+ Group, which includes stores such as B&Q, Habitat, and Great Mills, sells products labeled as coming from

Vietnam, but probably smuggled out of Cambodia. In December 2002 the prime minister of Cambodia said he would expel Global Witness, an organization that has helped publicize deforestation and human rights violations, for "abus[ing] our national sovereignty, our political rights and inflict[ing] damage to our reputation."[24] Unfortunately, he did not similarly fulminate against transnational timber corporations that abuse national sovereignty and political rights and inflict damage to reputations. Some things are better left unsaid.

Or we can talk about Siberia. The breakup of the Soviet Union led to the collapse of the Russian economy and the arrival of carpetbaggers from around the world. Banks and "development" agencies such as the U.S. Overseas Private Investment Corporation and the Export-Import Banks of the U.S. and Japan provide infrastructure loans and political risk insurance to corporations bringing American-made logging equipment into Russia, and exporting raw Russian logs to Japan, China, and Korea. Korean and Malaysian conglomerates now control huge concessions in the Russian boreal forests, and China smuggles Russian logs across the border.

All over the world, forests fall.

Globalization in the Real World

> I see in the near future a crisis approaching that unnerves me and causes me to tremble for the safety of my country. . . . Corporations have been enthroned and an era of corruption in high places will follow, and the money power of the country will endeavor to prolong its reign by working upon the prejudices of the people until all wealth is aggregated in a few hands and the Republic is destroyed.[1]
>
> *President Abraham Lincoln*

The global trade in wood and paper is no more sustainable than the global trade in human or nonhuman animals (both of which are also booming under the global economy). So often when I read about globalization in the newspaper or hear about it on television, the words and phrases quickly become so much jargon, so much babble. Structural adjustment, GATT, FTAA, NAFTA, WTO, IMF, free trade (if it's really free, why do they put sanctions on those who don't want to participate, and use police to violently eliminate protests by those who oppose it? If it's really free, why can't we opt out?). Hearing their voices, I quickly lose track of what they're saying, and of the relationship between globalization, impoverishment of human communities, and the despoliation of the natural world. But I guess that was the point all along.

So we'd like to provide a brief overview of the mechanisms and dynamics of globalization, especially as they pertain to the ongoing destruction of the world's forests.

The consuming elites (that is, the middle and upper classes of the United States, Europe, and Japan, and to some degree the upper classes in every other country) have an insatiable demand for luxury goods, commodities, and consumer products, including wood and paper products. I'm sure you can see the problems with infinite demands on a finite planet.

The Northern elites' governors and businessmen use debt schemes, bribery, weapons deals, and other unethical and illegal methods to get the Southern elites to collude in giving transnational corporations access to peoples' land, allowing (or encouraging) corporations to run amok—to cut down shrinking forests, to use dysfunctional industrial methods that pollute and waste, and to replace forests with failing plantations.

From the perspective of those in power in the North, if Southern elites aren't enticed by the crumbs of luxury and provide a massive flow of resources, then you might drive domestic plywood and paper manufacturing facilities out of business with a flood of underpriced panels and paper. If that doesn't do it, you can get the World Trade Organization tribunal to threaten economic sanctions for some imagined or trifling violation of international trade regulations. If that doesn't work, you can declare an embargo against the import of foodstuffs. And of course gunboats always wait just beyond the horizon. . . .

It's all just good old-fashioned colonialism, which my Webster's dictionary defines as "(a) control by one power over a dependent area or people; (b) a policy advocating or based on such control." And it's no coincidence that the rich of the world still control the colonies—although few are so honest or undiplomatic as to call them that—because many of the colonial structures were simply left in place after "independence." Corporate access to land, resources, and markets; debt peonage; tax structures favorable to those in power; commodity pricing aimed at driving small producers off their land; and the massive export of resources remain in the same mold as five hundred years ago. Only the names describing these mechanisms—and the names of those in power—have changed. In some countries, poverty is much worse than it was under direct colonial rule. The surviving forests themselves are in fragments.

Listen to this voice from the real world, of evicted Brazilian peasant Lazaro Correia da Silva: "Now I'm living in limbo, on the edge of the world. I've lost my land, now I have to work on the ranches. I work and they don't pay me. And then there's the police here. We go to them and try to get our money from the ranches and they put us in jail. I don't know what I'm going to do."[2]

The financial, legal, and political structures and mechanisms of globalization are woven into a complicated web manipulated by financial and technical experts on behalf of political and economic elites. This web usually obscures the real foundation that lies underneath: police and military power. We could give you textbook definitions of these structures and mechanisms, but the textbooks are generally written by these same experts, and the definitions often hide what's actually happening: people are being dispossessed of their commonwealth, as this commonwealth is turned into wasted commodities extracted for manufacture into luxuries for the elites. This puts us all—writers and readers—into a hard space, because just as basic knowledge of fire ecology can help inoculate readers against lies perpetrated by the timber industry and their supporters, a knowledge of terms used by those to promote globalization might help inoculate readers against the lies put forward by this particular group. But the terms are intentionally confusing and inevitably dull, so it is with trepidation, then, that we push forward with some definitions of the most commonly used mechanisms of globalization, in order to understand what happens in the real world, to real people, to real landscapes.

Before we do that, however, we'd like to bring in another voice from the real world, this from a Kayan native from the Uma

Bawang longhouse in Baram, Sarawak, Malaysia: "When I think of our land which is destroyed by the kompeni, it really pains in my body now. Now we can't find wood for our boats. The only wood left are the logs going down the river, there is none left on the land. They just bulldoze across our lands, now it is only sand and stones. Is it right for them to do this? What is the meaning of this? This is my land, my fruit trees. Yet they ask the polis to arrest me."[3]

The first set of terms has to do with financial mechanisms of globalization, and consists of such gobbledygook as debt instruments; structural adjustment; foreign direct investment; foreign aid; multilateral development programs; the transfer of destructive technologies; undervaluation of natural forests; subsidy of unsustainable practices; tariffs, taxes, and credit structures; transfer pricing; price fixing; control of resource inventories, customs houses, and national accounts; using timber taxes and royalties for general government revenues (or consumption by the elite) rather than for forestry programs; and various methods of keeping economic benefits within the political and business classes.

Let's take these one by one.

Debt instruments and "development." Corporations and government agencies loan money to "poor" countries. This money is used to purchase equipment and expertise from the "rich" country's corporations. "Development" projects focus on infrastructure necessary for the extraction and export of the poor country's commonwealth. The poor then have to pay back the loan. They pay for their own deforestation, have to live with the devastation, and pay interest to boot.

Structural adjustment programs. I'm sure you've read in the newspaper that this or that country is undergoing "structural adjustment." For the longest time I never understood what that meant: Are the World Bank and the International Monetary

Fund (IMF) sending over planeloads of chiropractors to parachute into troubled cities? But here's what it is: When loan payments mentioned in the previous paragraph become too great a burden, the World Bank and IMF step in to restructure (bankrupt) the poor nation's economy. Public services, including schools and hospitals, are eliminated or sold to corporations for a little cash, the prices of basic food and energy are raised, and an increasing amount of the gross domestic product is used to repay ever-higher debt loads. The cure for staggering debt is more debt. This makes those in power smile, because it guarantees greater access to the forests and other raw materials of the poor nation. For example, Ghana's structural adjustment program of the mid-1980s raised timber from 3 to 8 percent of the country's gross domestic product, and to 14 percent of its exports (number three after cocoa and gold).[4]

Foreign direct investment. Investment is a good thing, right? If you're a businessperson, you want people to invest in your business, right? Well, that depends on what they want from you. We saw early in this book—and in the redwood forests of California—what happened when Hurwitz "invested" in Pacific Lumber. Similar things happened when Boise Cascade "invested" in Mexico, Freeport McMoRan "invested" in Irian Jaya, Weyerhaeuser "invested" in Indonesia, and so on. "Foreign direct investment" is a euphemism for (1) offering huge loans to buy the creditor corporation's goods and services, and (2) buying out the poor nation's businesses and resources. The result of foreign direct investment is the extraction of resources and the destruction of communities.

Foreign aid. What could be better than investment? Direct aid, of course. I'm sure you've heard people complain about the ingratitude of poor people who hate the United States even as it hands over aid to them. Unfortunately, "foreign aid" is a euphemism for (1) offering loans to buy the rich nation's corporations' goods and services (see above), and (2) giving money to the rich nation's corporations to penetrate the poor nation's markets. For example, when the United States extends foreign aid to Russia, a percentage of the loan is earmarked for the purchase of, say, Caterpillar tractors manufactured in the United States. In other words, the U.S. government gives money to the Caterpillar corporation, and Russia pays it back with interest to the United States.[5]

Multilateral development programs. Another way to say multilateral is "ganging up." These programs happen when several rich nations combine to offer loans to a country that has, from the perspective of the rich, too many undeveloped resources and too many public services. It is often done under the auspices of an international agency such as the World Bank (of which the United States is the dominant lender) or a regional agency such as the Asian Development Bank (of which Japan is the dominant lender).

Technology transfer. When the elites talk about the transfer of destructive technologies, they usually leave off the word *destructive,* and call it technology transfer, as if they were giving away something valuable. Here, nonindustrialized countries are often given (more often, loaned at high prices) mechanized road-building, logging, and milling equipment to facilitate the rapid exploitation of remote forests.

Undervaluation of natural forests. Here, costs are ignorantly (or more likely deliberately) externalized—left for others to pay. They are transferred onto future forest productivity, nonhuman species, forest dwellers, and future generations. The undervaluation of the natural world and the externalization of as many costs as possible are both central to our culture's economy. Without them the "profits" of an extractive economy would be impossible.

Subsidy of unsustainable practices. Of course the economy would collapse immediately without massive subsidies. Just a few of the many subsidies for forest-destroying corporations include: building logging roads into forests at public expense; publicly maintained transportation and port facilities for the export of wood and paper products; tax breaks for giant pulp and paper mills that consume entire forests; use of public tax monies to restore degraded forestland; and publicly funded unemployment benefits and job training for workers laid off after their employer cuts and runs.

Tariffs, taxes, and credit structures. Charging exorbitant interest rates, demanding natural resources in payment of debt, or demanding the elimination of tariffs and other taxes that protect domestic industries and resources from outside competition are three different flavors of the same scheme: Defrauding the poor.

Transfer pricing. This is a fancy name for the routine practice of undervaluing transactions (usually between affiliated companies) in order to evade taxes and overcharge customers.

Price fixing. "Competing" firms sometimes agree to set prices too high (to gouge consumers) or too low (to rip off suppliers). Among politicians and businesspeople in the United States, this is normally called *competition*.

An important tool used by those determined to extract resources from a poor country is the **control of resource inventories**. Foreign corporations manage to buy or extort complete oversight over forestlands, including knowing how much timber is left and how fast it is being cut. By now there can be little question about how far we can trust the heads of corporations to give honest accountings of these forests (or anything else) and how fast they're being destroyed. Worse, foreign "aid" loans are used to inventory and extract timber—meaning, as always, the country pays twice to deforest its own landbase. Likewise, rich nations have often seized poor nations' **customs houses and national accounts** in order to use their income as debt payments. This violation of sovereignty—what's that?—is commonplace. Further, the World Bank and International Monetary Fund enforce restructuring of national economies and government spending in order to give debt payment priority over social services, environmental protection, and investment. In fact, over everything. Imagine if Iraq attempted to dictate the U.S. budgetary process. American citizens would be up in arms, because the right to determine the U.S. budget is generally reserved for transnational corporations.

Yet another way those in power control resources is by convincing the local elite to **use resource taxes and royalties for general government revenues** (or consumption by the elite) rather than for reforestation. This of course means no money goes into long-term forestry programs, which from the perspective of those in power is a fine thing: By the time the forests are gone, those in charge are ready to deforest another country.

Some of the other methods used to keep economic benefits within the political and business elite include (1) allowing foreign aid monies to be siphoned off by government officials, sometimes sending them directly to the dictators' Swiss bank accounts for safekeeping (note that these monies must still be repaid by the country's taxpayers); (2) maintaining tariffs (which is a fancy word for taxes) on imports of basic foods and energy, while subsidizing luxury goods for the elite; and (3) selling weapons and training the poor nation's police and military to maintain order while the country's resources are looted. This is more than commonplace. Exxon funds the Indonesian military. Shell Oil funds Nigeria's military. British Petroleum and the U.S. government fund the Colombian military. In return, these militaries protect oil pipelines and quell protests.

Those are some of the financial mechanisms by which the rich steal from the poor in the global economy.

Here is what happens when someone opposes the destruction of forests. Raul Zapatos lives in the Philippines. He was the team leader of a Department of Environmental and Natural Resources (DENR) strike force, and thus responsible for stopping timber theft. Although the DENR is, not surprisingly, corrupt, Raul was not. In 1989 he twice stopped a truck full of illegal timber. The second time, he refused to release it, even though it was being used by the mayor, whom he considered a friend. On January 14, 1990, Raul was asleep at the strike force headquarters when the mayor, his bodyguard, and the police attacked with rifles and grenades. Raul returned fire with an M16, killing the mayor and wounding his bodyguard. For this he was found guilty of murder and frustrated murder, and sentenced to life imprisonment.[6]

Next, it's time to describe some of the legal mechanisms, including such "acceptable" mechanisms as land "reform" programs and property laws, responsible for maintaining inequitable ownership and labor relations. These also include legal and illegal monopoly rents, tax evasion, price fixing, and bribery.

We've already mentioned some forms of "reform," which include resettlement, colonization, and so on. The key thing to remember about land programs is that when a government promises new land to someone, it has to take the land from someone else. And you *know* that "someone else" ain't gonna be the rich. Nor the foreign corporations. Nor the leader's friends. Most likely it's going to be the indigenous. And frankly, the offer is nearly always an excuse to then give it, not to the poor to whom it was promised, but to their fellow elites.

One of the most invisible mechanisms is that of **property laws,** such as the private ownership of land. Private land ownership is a very strange concept: If I own no land, I do not have the right to sleep on this earth without paying someone. I need to pay rent simply to exist. European notions of property law have been extended across the world, often by force. Those who have refused to recognize these notions—believing instead that the land is alive, that it is held communally, can never be bought and sold, and must be used (insofar is it is "used" at all) for the benefit of the whole community, including the landbase itself—

have had their lands confiscated. Those who have resisted have been killed.

Once you've put in place the European notion of private property ownership, you have in place the notion of rent, and further the notion of monopoly rent, which arises when one person or corporation owns exclusive control over some unique item. This unique item can be housing, but it can also be access to a forest. It can be access to water. It can be access to medicines. It can be access to food. All of these resources, and nearly all other necessities of life, are increasingly controlled by large corporations. When this or that unique item is necessary for life, monopoly rents skyrocket. Profits soar, making those who run corporations feel what passes for happiness in our culture.

Tax evasion is a fairly common means by which those in power defraud the poor. This is true in the United States, where dozens of wood and paper companies pay *less* than nothing for taxes, but instead actually receive taxpayer money. It is true the world over. In Cameroon, Ghana, Indonesia, and many other countries, a third or more of the timber cut is never reported, and thus never taxed. The fine for illegal logging in Ghana is less than the price of one cubic meter of wood.[7]

Price fixing occurs at nearly every step of the extraction process in the global economy. Logging equipment is overpriced when it is sold to the poor. Extracted timber is then routinely underpriced. Consumers are charged monopoly rents. It's a racket.

Bribery takes many forms. In the third world, sometimes it's called speed money: Bureaucrats don't provide necessary licenses and paperwork unless you bribe them. In the United States, the preferred term is *political contributions*. Depending on the culture and circumstance, bribery is sometimes illegal and sometimes acceptable in the form of gifts or rebates. The point is that some can afford to make the payments and others can't.

• • • •

Yet another report from the real world, this from Liberia, entitled, "Logging Companies as Conduits for Domestic Political Repression." It states, "Logging companies now constitute the most powerful and politically insulated layer of our national bureaucracy. Logging companies' private armed militias have now replaced our national police apparatus in rural Liberia. . . . Notorious for human rights abuses and the commission of brutal atrocities against innocent civilians are the Oriental Timber Company, OTC, operated by Gus Konhavowen, a Dutch national, and the Inland Timber Company, ILC, owned by Charles Taylor [President of Liberia] and managed by two historically criminal brothers, General Oscar Cooper and Maurice Cooper. . . . On direct orders of General Oscar Cooper, members of the ILC militia consistently and forcibly enter private homes of citizens at night and rape women of individuals who publicly criticize Taylor's logging policies. The ILC-controlled militia has also gained national notoriety for public floggings, arbitrary arrests, and widespread torture of peaceful citizens kept in illegal militia detention. . . . To compel disgruntled villagers to work for them, tribal people are militarily prevented by logging companies from engaging in subsistence farming activities on their own land. The logging companies' official justification is that the Forestry Development Authority, FDA, headed by Charles Taylor's brother, Bob Taylor, has, through concession agreements, assigned those forests to them and that inhabitants no longer have dominion over them or are permitted to undertake farming activities in such areas without the expressed approval of loggers holding such concession agreements. Conversely, however, the logging companies discriminately do allow their workers and relatives of those workers to farm in the very areas they claim to be off limits to citizens who are at variance to their logging practices. Undoubtedly, opposing a logging company in Liberia under Charles Taylor constitutes a total denial of one's personal livelihood as a human being and as a citizen."[8]

Which brings us, finally, to political mechanisms, webs of patron-client relationships that tie together the political, bureaucratic, military, and business elites within and between the producing and consuming nations. These webs include brainwashing that is euphemistically called public relations, phony public participation programs, underpaying civil servants to facilitate their corruption, repressing and eliminating resistance by peasant and native peoples, gunboat diplomacy, and engineering coups to install client regimes.

Corporations and governments routinely hire **public relations** firms to spin their atrocities into gold. Entire books have been written on this, notably *Toxic Sludge is Good For You,* by John Stauber and Sheldon Rampton. An example concerning forests is Chlopak, Leonard, Schechter & Associates, a PR firm that specializes in crisis management. It helps oil corporations fend off environmentalist and human rights groups opposing a 400-mile-long pipeline in Peru that will pass through indigenous homelands in the Amazon rainforest.[9]

Public participation programs are replacements for genuine democracy. Public relations firms, government agencies, and consensus groups have developed sophisticated techniques for confusing, exhausting, and co-opting the concerned public. The fundamental understanding that guides this participation—and those in power are exquisitely aware of this understanding—is the impossibility of true negotiation between parties of grossly unequal power. For example, all through my years of filing timber-sale appeals, I participated in countless meetings with Forest Service officials who screwed encouraging looks to their faces and politely pretended to listen to our comments, which both we and they knew would be ignored. I also wrote out god-only-knows how many comments, which were duly noted, but also ignored. But I did participate in public process! Is this a great country, or what?

My experience is not unique. That *is* public participation in this country. We are "allowed" to "speak truth to power" all we want, but everyone knows that those in power will ignore these truths and go ahead and do whatever the hell they want. It's far more efficient to let the people have their meaningless say than it is to trundle them off to some distant gulag. This way we can all happily pretend the system works. Unfortunately, the system *does* work—in fact all too well—but never the way we were told.

Especially in the third world, **civil servants are routinely underpaid.** This facilitates their willingness to accept bribes to not prevent the theft of resources. (If Forest Service acquiescence in timber theft in the United States is any indication, American civil servants do not even need bribes to facilitate corruption.) A potent third-world example comes from Cameroon, where in a classic "structural adjustment" in 1992 and 1993, civil servant salaries were slashed 60 to 70 percent, making corruption of those servants an easier task. The result? Less than a third of Cameroon's timber is locally processed, a third of the cut is never even declared, and two-thirds of the timber taxes are never collected.[10]

Central to the theft of resources are the **military and police repression** of peasant and native peoples, gunboat diplomacy, and coups to install client regimes. This is, indeed, a primary function of the U.S. military. As former Secretary of Defense William Cohen said to a group of Fortune 500 leaders, "Business follows the flag. . . . We provide the security. You provide the investment."[11] As shown, business doesn't even have to provide the investment: That's borne by the public.

The real-world results of these mechanisms are dispossession and poverty, forced labor, extraction of the commonwealth to produce raw materials and commodities, and the downward spiral of autonomous local markets into global economic production, consumption, depletion, and collapse. Collectively, these mechanisms are called globalization.

• • • •

There has been a lot of debate—much of it diversionary—for decades about the true causes of deforestation. Part of the confusion is that the immediate causes do vary by region, forest, and decade. Part of it is that the sociocultural-political-economic-ecological web of interdependence is genuinely complex and ever shifting. Yet another part is that confusion often supports the status quo. Additionally, but not accidentally, there's a lack of accurate information on the condition of the world's forests. Governments often do not keep good data, for a variety of reasons, including budgeting money for timber cutting but not for determining the actual conditions of forests, collusion between administrators and direct deforesters, and ultimately the fact that our culture doesn't value intact forests. Further, satellite photos often don't reveal the true extent of damage to forests. And there exist different definitions of nearly all of the terms: We are diverted by debating the semantics of deforestation, natural forest, old growth forest, frontier forest, sustainability, and so on.

Here is what we do know, summarized well by Peter Dauvergne in his book *Shadows in the Forest:* "Loggers irreparably decrease the economic, biological, and environmental value of old-growth forests. They also ignite the process of deforestation.... They build roads that provide access for slash-and-burn farmers. They leave debris and create open spaces that make forests susceptible to devastating fires. And they decrease the financial value of primary forests, providing incentives to convert logged areas (secondary forests) to commercial crops or large development projects.... Both direct and indirect factors and underlying forces contribute to rapid, careless, short-sighted logging. While... state managers and timber operators play direct roles, international corporations, markets, money, consumption, technology, and trade practices cast an oppressive shadow that constrains... decisions, provides incentives for quick and destructive logging, and accelerates deforestation."[12]

All that said, it is possible to determine the causes of deforestation in many regions. Africa provides an example of the wide range of reasons forests fall. In Gabon, Cameroon (where timber is the number-one source of foreign exchange revenue), the Central African Republic, the Congo, and Equatorial Guinea, a primary cause is legal and illegal logging. In Nigeria (where Shell has made $30 billion from Ogoni lands since 1958), Ghana, Madagascar, and Tanzania, a primary cause is oil and mining. In Ivory Coast, Ghana, Nigeria, Cameroon, South Africa, and the Congo, tree plantations—rubber, pine, eucalyptus, acacia, cypress, and oil palm—are a central cause of deforestation (remember, plantations are not forests). In Uganda, the Norwegian corporation Treefarms has evicted 8,000 people from thirteen villages and planted pine and eucalyptus as a "carbon sequestration" scheme to fight global warming. In Nigeria, Madagascar, Tanzania, and Senegal, shrimp farming destroys coastal mangrove forests. In Uganda, dams have drowned forests. The forests of Liberia, Angola, and the Congo are or have recently been devastated by war. In the Congo, for example, combatants have taken to systematically raping, torturing, killing, *and eating* indigenous pygmy forest people, including children. (These actions are called by the perpetrators a form of "vaccination" to rid the country of "a disease"; indeed, the code name for this operation is "Wipe the Slate Clean.") Liberia, France, Belgium, and other nations, as well as corporations, exchanged weapons for gold, diamonds, and timber. Illegal timber exports reached $53 million per year during the Liberian civil war of the early 1990s. Carpetbagging corporations routinely move in after war to take advantage of the shambles. In Liberia these corporations include LAMCO (USA-Sweden), Bridgestone (Japan), and Oriental Timber (Malaysia).[13]

Using the best current estimates, which come from the World Resources Institute, we can summarize that three-quarters of the remaining frontier forest in Africa is under threat, 79 percent of that by logging, 12 percent by mining, roadbuilding, and other infrastructure, and 17 percent by agricultural clearing. (These percentages add up to more than 100 percent because some forests face multiple threats.)[14]

Sixty percent of the frontier forests in Asia are under threat, half of that by logging, 10 percent by mining and roads, and 20 percent by agricultural clearing.

Twenty-six percent of North American native forests are under imminent threat, with the primary dangers being logging (84 percent) and mining and roadbuilding (27 percent).

Eighty-seven percent of Central American forests are threatened, slightly over half of that percentage by logging, 17 percent by mining and roadbuilding, and 23 percent by agricultural clearing.

Fifty-four percent of South American forests are under direct threat, with 69 percent of those forests being threatened by logging, 53 percent by mining and roadbuilding, and, despite what we've heard in the corporate press about peasants destroying the forests of the Amazon, only 32 percent by agricultural clearing.

Because of the remoteness of the Russian Far East, only 19 percent of the native forests of Russia are threatened, 86 percent of those by logging, 51 percent by mining and roadbuilding, and 4 percent from agricultural clearing.

All of the forests of Europe are under threat, with 80 percent of that threat from logging.

Over three-quarters of the native forests of Oceania are under immediate threat, 42 percent of these by logging, 25 percent by mining and roadbuilding, and 15 percent by agricultural clearing.

Worldwide, almost 72 percent of the threat to forests is from logging, 38 percent is from mining and roadbuilding, and 20 percent is from agricultural clearing.

The exact percentages can be endlessly debated, but the reality exists.

• • • •

Globalization. Structural adjustment. Deforestation. Big latinate words. We've tried to define them. But what do they mean? Here is yet another story from the real world: "Teak is the second largest legal money-maker for the [Burmese military regime] SLORC. . . . Burmese and Thai loggers use elephants to move logs around, drugging the animals with large amounts of amphetamines, to which they can become addicted. Many elephants get sick and die because of overwork due to the pressure to log teak at ever faster rates."[15]

The processes of globalization we've described so far can be loosely summarized by a sequence of motivations and actions. Insatiable consumption leads those in power to invade other nations to steal their trees. This leads to the conversion of forestland to cropland, which leads to former forest dwellers being forced to work as farmers or factory laborers. This leads to poverty, which leads to refugees entering the receding frontier forests for food and fuel. All of this leads to international debt, which leads to agribusiness and raw materials exports, which leads to more debt and greater poverty. This finally leads to collapsing domestic economies, structural adjustment, and acceleration of the whole cycle.

Consuming the World

> We have never said we were on a sustained-yield program, and we have never been on a sustained-yield program. Let's get to the heart of it. Sure, it's extensively logged, but what is wrong with that?[1]
>
> *Bill Parsons, Rocky Mountain Regional Operations Director, Plum Creek Timber Company, 1989*

> The world is in great danger. When the trees die, the Earth dies. We will be orphans without a home, lost in the chaos of the storm.[2]
>
> *Kayapo chief Tacuma*

North America, Europe, and Japan, with 17 percent of the world's population, consume three-fourths of the world's traded timber. The United States is by far the largest consumer of wood and paper products. The global wood and paper industry manufactured $850 billion worth of products in 2000, $260 billion worth from the United States.[3] Of the major categories of wood products (fuelwood, roundwood, lumber, panels, pulp, and paper), the United States is the top consumer of all but fuelwood. And in every one of those categories, the United States consumes more than twice as much as the next-largest consuming nation. With less than 5 percent of the world's population, the United States consumes between 25 and 38 percent of the world's wood and paper products.[4] To facilitate this overconsumption, U.S. wood products corporations are major actors in other nation's forests.

Canada forms a good case study, one that is sadly repeated the world over. In 2000, the United States got 40 percent (or 6.6 million tons) of its newsprint from Canada.[5] The United States also gets a third of its lumber from Canada, primarily British Columbia.[6] At least ten of the top fifteen wood products corporations based in B.C. sell more than half of their products outside of Canada, primarily to the United States.

Weyerhaeuser, Pope & Talbot, Louisiana-Pacific, Champion International, and other U.S. companies are among the top recipients of timber quotas in Canada. Furthermore, U.S. corporations and investors control interests in several "Canadian" corporations: West Fraser Timber is controlled by the Ketcham family of the United States, and Cariboo Pulp and Paper is owned by Weldwood, which is owned by Champion International (now part of International Paper).[7] But the real U.S. giant in Canada is Weyerhaeuser. Not only did it recently acquire MacMillan Bloedel, an immense Canadian company in the process of deforesting much of British Columbia, but it holds 34 million acres of long-term licenses to timberland in Saskatchewan, British Columbia, Alberta, and Ontario.

Ninety percent of the lumber manufactured in British Columbia is exported, with two-thirds of it going to the United States.[8] Seventeen percent of B.C. lumber goes to Japan—mainly structural lumber cut from coastal old-growth forests.[9] The B.C. government provides timber corporations with below-cost woodcutting rights (costing Canadian taxpayers an estimated $2.8 billion per year) and direct taxpayer bailout of antiquated and polluting mills ($329 million just for the Skeena Cellulose mill). It also waives environmental protection laws ($950 million annually) and allows (or rather encourages) timber companies to ignore aboriginal title to forestland (requiring hundreds of millions of dollars in compensation).[10]

When environmentalists succeed in temporarily halting some deforestation, you'll often see articles in the corporate press mentioning the number of houses that could have been built were environmentalists (or frequently "environmental extremists") not so selfishly intransigent. But what these articles nearly always fail to mention is that more than a third of the trees cut are pulped for paper.[11]

The truth is that the pulp and paper sector drives the entire wood products industry. Pulp mills are extraordinarily expensive, and so when they're built—usually with massive subsidies of public money—they may as well be huge. The size of the individual mills plus the incentive of free public monies leads inevitably to overcapacity: The capacity to pulp more wood than even a further-subsidized market can bear. But the mills aren't generally *entirely* subsidized, and so the corporations that build them end up with a debt. This means the mills have to be run, whether or not there's a market for the paper, which means trees "need" to be cut. Lots of them. The trees that aren't made into paper are made into lumber or panels, or burned as fuel for the mill. One consequence of this is that the market becomes glutted with lumber, plywood, and so on—mere by-products of the pulp and paper process.

Another consequence is that more forests are destroyed. Before wood fiber is pulped, whole trees are chipped into tiny pieces at chip mills. Because they use entire trees, including small ones, chip mills encourage clearcutting and short cutting cycles, often consuming entire forests a hundred miles or more from the mill. Last year alone, more than a million acres were cleared to feed 140 chip mills in the southeastern United States.[12] And the number of chip mills is increasing: in less than twenty years in North Carolina, for example, the number has jumped from two to seventeen. These seventeen chip mills alone can process 250,000 tons of chips per year: 8,000 to 12,000 acres of hardwood and softwood trees.

The same processes occur, constantly, around the world.

• • • •

Just so we're clear that we as a culture are moving in entirely the wrong direction, we need to note that U.S. paper consumption increased fivefold between 1920 and 1990. World paper consumption increased from 15 million tons in 1910 to 463 million tons in 1996.[13] The global South and Eastern Europe, with 84 percent of the world's population, consumed less than a quarter of the paper worldwide, while the global North and the Asian "tiger" nations, with 16 percent of the population, consumed over three-fourths of the paper. The 460 million people in the United States, Japan, and Germany consumed almost half of the world's paper; the United States alone consumed more than Japan, China, Germany, and the UK combined.[14]

Hundreds of Dayak, Penan, and Iban people have been arrested, jailed, and beaten for erecting blockades to stop roadbuilding and logging in Sarawak. They say, "We will defend our land at all cost. We will never surrender it. We have our dignity and we must leave something for our future generations."[15]

What are we doing? Why are we deforesting the planet? How much longer until we, like the people of the forests, wake up and begin to fight back?

We can no longer plead ignorance. To be truthful, that has never been an option. Apathy, perhaps. Ignorance, never. The effects of deforestation have been known since ancient times. Empires from the beginning of civilization to the present have expanded to acquire wood supplies for shipbuilding and fuel for industry, and have collapsed when those wood supplies were depleted.[16]

Yet the global timber industry keeps trying to escape the ecological limits to raw materials and the social and economic limits to markets, by relying on frontiers. They cut the forests of New England, then the Midwest, then the Pacific Northwest, then Canada. . . . (It's as Jude White of International Paper said to forester Gordon Robinson, "Hell, Robbie. We're on sustained yield. When we clean up the timber in the West, we'll return to New England, where the industry began.")[17] While multinational timber corporations use the rhetoric of sustainability and jobs, the millennia-old reality of cut and run continues. In the words of lumberman Henry F. Chaney, timber towns are "never considered anything but tools in the rescuing [sic] of the timber and would be discarded just like a worn-out hoe or plow or any other piece of equipment." He continued, "Outside the rescuing [sic] a timber body in some pestilential swamp in Louisiana or pine flat in Wisconsin or Michigan, what other purpose would be served by maintaining a town there?"[18]

Despite the public-relations strategies of corporations, timber industry overcutting has been confirmed by numerous industry, academic, and government studies. Now that the end of the forest frontier has been reached, free trade agreements are threatening to remove the last barriers to total industrialization and depletion.

This all brings us back to globalization, a word we've discussed through some of its characteristic processes. Here we're going to define it anew: *Globalization* is a current term for the horizontal and vertical integration of manufacturing and trade on an international level. Vertical integration is where one company controls processes from beginning to end of extraction to manufacturing to distribution and sales. The vertical integration of the wood products industry has been called by one expert "probably the single most recognized characteristic of the industry,"[19] meaning, for example, that most paper sales are by corporations which also control timberland. Now the horizontal integration of the industry is also being completed, as corporations like International Paper spread their operations to dozens of countries. In other words, fewer and fewer corporations control more and more of the world.

But you knew that.

All of this creates problems, just as the spread of cancer throughout a body "creates problems" for the host. These problems are not merely ecological, but economic as well. Below we discuss a few such problems.

Globalization forces everyone to compete with the cheapest producers.

In the early 1990s, bleached hardwood pulp cost $78 per ton to produce in Brazil, $156 per ton to produce in eastern Canada, and $199 per ton to produce in Sweden.[20] Of course the primary reason pulp was cheaper in Brazil was that more of the costs could be externalized—sloughed off onto the natural world, onto poorly paid workers, onto members of local communities, onto future generations. Following the logic and reward system of capitalism, paper producers shift to these places of greatest externalities, and workers and governments are drawn into competition to see who can destroy their own communities and land most quickly and cheaply.

Corporations, aided by their governments' agricultural and foreign aid policies, regularly dump products into overseas markets at prices below their real cost.

This destroys local economies and self-sufficiency, and further promotes the externalization of costs. For example, since joining GATT in 1986, and especially since the passage of NAFTA in 1994, tariffs on Mexican imports of pulp and paper have decreased. As a result, the Mexican paper market has been glutted by imports of U.S. paper and by the entry of global manufacturing corporations like Weyerhaeuser and Smurfit-Stone. Paper prices collapsed, as did the Mexican paper industry, leaving the global corporations free to take over the market.

The interconnected nature of the global economy puts particular regions at risk.

When Asian economies collapsed in 1997, regional timber economies in the American Northwest and South were hurt. When Columbia Forest Products closed its hardwood veneer plant, 225 workers lost their jobs. Michael King, an attorney representing Columbia, "listed a number of factors that conspired against the plant, none of them related to the work force or management. 'It's part of the whole global economy sort of thing.'"[21] Globalization destroys jobs in several ways. Jobs are shifted overseas as corporations seek cheap resources, cheap wood, government handouts, and foreign markets. Jobs are also lost in an absolute sense, as hyper-competition causes consolidation and sell-off of facilities. This is especially marked in the pulp and paper and panel industries, where excess capacity has been a long-standing problem.

Capital-intensive manufacturing (such as pulp and paper) destroys local small-scale ecological and economic systems.

It also reduces employment, as technology and mass production replaces human labor. World-class pulp mills, which can cost $1 billion to build, create additional debt in poor countries—and can require $1 million in capital investment *per job*.[22] Of course the locals would be far better off if someone handed them a half million dollars each and told them to do what is best for the forests. The capital required for the industry to continue growing has perhaps hit its limit, and corporate spending is increasingly going toward consolidation—buying out competitors rather than building new capacity.[23] All of this helps the biggest corporations, not small companies, and certainly not workers. As one writer puts it: "It has been more than a quarter century since the primary producers in the paper industry created a new job."[24]

Mergers and acquisitions destroy forests as well as jobs.

When timberland is sold, the buyer's debt (which is the seller's windfall) often forces quick liquidation of timber assets, the closure of mills, and the layoff of workers. Examples include James Goldsmith's takeover and dismantling of Crown Zellerbach, Georgia-Pacific's takeover of Great Northern Nekoosa in Maine, and MAXXAM's aforementioned buyout of Pacific Lumber.

Paper manufacturing capacity far exceeds consumption.

As prices slump, investment in huge paper-making machinery becomes possible only for the largest corporations, and even they require larger and larger subsidies from governments in the form of low-interest loans, tax breaks, and infrastructure subsidies. Further consolidation of the industry, excess capacity and consumption, and resource depletion are encouraged, to cover the costs of running these mills.

Boom and bust cycles are exacerbated.

The cyclical nature of the paper and wood products industries is caused by fluctuations in inflation, currency and interest rates, demand, and supply. Every five to ten years the downside of the cycle leads to price drops, wage cuts, mill closures, and ever tighter market control by corporations large enough to survive the recession. Traditional overcapacity in the timber industry exacerbates economic cycles, and results in the continuing consolidation of regional industries into a global industry dominated by a few huge international corporations, as less efficient (or less subsidized) producers go out of business. Five corporations now produce 60 percent of all pulp in Japan, and five corporations produce 85 percent of the newsprint in Europe.[25] Alternatively, in boom times, lumber and pulp prices soar, easy credit becomes available to build greater capacity, and then the inevitable next collapse makes debt harder or impossible to pay off. As the industry loses money (the Canadian pulp industry lost $4 billion from 1991 to 1993), cost-cutting leads to layoffs and wage cuts, reduced environmental protection, and delayed maintenance. The U.S. paper industry permanently lost 6,000 jobs during the 1990–91 recession alone, and the current wave of consolidation in the pulp and paper industry is expected to cost another 50,000 jobs.[26]

Mechanization, raw material exports, and overcutting destroy jobs.

Corporations make workers and environmentalists into scapegoats—remember "jobs versus owls"—while engaging in wage cuts and routine violations of environmental and health and safety protections. In reality, jobs are lost through overcutting caused by overconsumption (encouraged by advertising), mechanization (called "increases in productivity"), and the wholesale export of raw logs, wood chips, and pulp. We've already discussed how both paper and lumber production have gone up over the past few decades, while the number of jobs has gone down. The numbers of jobs permanently lost never reveals, however, the full extent of the human damage, because it fails to include wage and benefit cuts, increase in part-time and contract jobs, and routine "temporary" layoffs. It's important to remember that a corporation's "increase in productivity" is a human being's unemployment, and a forest's death.[27]

Globalization shifts the income from workers to investors, and shifts the costs from investors to communities.[28]

A week after the merger of Jefferson Smurfit and Stone Container, to provide a typical example, the new Smurfit-Stone said it would pare its work force by up to 3,600 jobs, or 10 percent.

"The restructuring is a difficult action to take, considering the impact it will have on jobs and on communities," said Roger W. Stone, president and chief executive. "However, it contributes significantly to achieving our synergy goals, increasing our competitive edge and building shareholder value."[29] Meanwhile, when International Paper and Union Camp announced their merger, a spokesman said closings are possible but that no decisions had been made. An International Paper manager observed, however, that "[Job losses are] a regrettable situation. However, we live in a free market economy."[30]

His statement is nonsense, for several reasons. First, "the market" has always been driven by politically motivated and anti-competitive subsidies, which means "the market" is in no way "free." Second, those who oppose the theft of their resources are co-opted, displaced, or even killed, meaning, once again, "the market" is not "free." Third, his statement imparts an immutability to this mythical free market, as though it were an inescapable fact of life, like the existence of gravity. Yet the dominant economic system is the result of numerous choices, many of them very bad for most on this planet. We should never forget that.

Yet another report from the real world says: "According to officials of the Brazilian government's Indian Foundation, there is now scarcely any mahogany being sold by Brazilian companies which does not come from [Indian] reserves. Having exhausted stocks in nearly all the places in which they [the mahogany] could be legally exploited, the big timber cutters have swept into the protected areas set aside for Indians or wildlife. Most of the reserves in both the eastern and southwestern Amazon are being exploited by timber cutters, to devastating effect.

"Half of the 1,200 Uru Eu Wau Wau Indians of the state of Rondonia have been killed since 1981, either by timber cutters' posses or by diseases introduced by the invaders. Two sawmills are working inside their reserve, and the mahogany that one of them is extracting moves—through a series of middlemen and a British importer—to the furniture restoration departments in Buckingham Palace and Sandringham.

"The Korubu of the Javari Valley in Amazonas state, have, like the Guapore, fled from outsiders, but they too are now being forced into violent contact. Three months ago two members of a logging team failed to return to base. Their disappearance was blamed on the Indians, and a posse was sent in, whose express intention was to eliminate all the Korubu they came into contact with.

"The Guapore, Uru Eu Wau Wau, and Korubu are among the many groups whose only contact with the timber cutters in their lands has been at the end of a gun, but perhaps the story of those who have struck deals with the loggers is even more tragic. A few years ago white men began to arrive in the territory of the Kayapo Indians, in the state of Para, in pick-up trucks laden with cheap merchandise: teeshirts, tinned food, radios, torches, and plastic toys. They handed them out to the Indians, who were glad to receive them.

"Several weeks later the same men turned up again, announced that they were timber cutters, that the goods they had distributed had been sold to the Kayapo on credit, and that they had come [like the IMF bankers!] to claim their debts. As the Indians had no money they [the timbermen] would collect what they said they were owed in the form of timber.

"The Kayapo could do nothing to stop them moving in and taking wood many hundreds of times the value of the merchandise they had distributed. Once there the cutters began to work on the more dominant members of the group, tempting them with promises of trucks, stereo systems, and prostitutes. Their behaviour has been compared by Indian rights campaigners in Brazil to that of drug pushers. Several of the village leaders became hooked, and started selling off the forest without consulting their people. The results for both the society and environment of the Kayapo have been catastrophic."[31]

Globalization feeds on itself.

Mergers and consolidation encourage more mergers and consolidation. Share prices jumped on the announcement of the merger of International Paper and Union Camp, and Wall Street analysts and traders gleefully anticipated it to cause even consolidation across the paper industry. "This really begins to put pressure on other paper companies to become serious about deal-making,"[32] said one analyst. Further consolidation in the industry is inevitable, according to another, who predicts that the number of U.S. containerboard mills (currently fifty to sixty) will be cut in half by the year 2005.[33] An industry journalist in Asia justified the restructuring by concluding that "being number two (or worse) at a time of overcapacity can be the kiss of death. . . . The process [of consolidation] will be a painful one for many . . . but a leaner, more efficient industry will emerge from crisis. . . . Rich Western companies have started to move in [to Asia]. Local M&A [mergers and acquisitions] is sure to follow."[34] The chairman of Smurfit-Stone, seemingly a winner in the corporate race to "survive," takes an even more enthusiastic view of the trend: "Undoubtedly, the increasing globalization of our customer base has also been a catalyst for the spate of mergers and amalgamations that are currently reshaping our industry. This most welcome process of industry consolidation will, I believe, secure a much brighter future for our businesses for the next two years and beyond."[35] A two-year planning horizon!

International trade destroys domestic economies.

Mexico makes a good example. Saying that Mexico was reaping the benefits of NAFTA, and the market was growing very strongly, Smurfit announced plans to spend $120 million to expand its corrugated and folding carton business in Tijuana, Mexico.[36] But the directors of Smurfit were lying. All is not well in Mexico. From 1995 to 1996 Mexican employment in wood products dropped 30 percent and production was cut in half, while consumption increased by 16 percent. Imports and the entry of foreign corporations have destroyed the Mexican paper industry, which had been using wastepaper for 75 percent of its fiber needs. The Mexican government's response has been to offer even more incentives to foreign corporations.[37]

Just as clearcuts are called "temporary meadows," global cut-and-run is called "innovative."

In 1995 *World Paper* wrote that "Diminishing access to forests, timber certification and recycling regulations have drawn innovative responses from our industry. . . . The search for fibre is moving offshore."[38] Let's translate: because timber corporations have deforested the region, and because local regulations are beginning to prohibit the further externalization of certain costs, timber corporations are leaving the country. They are doing what timber corporations have always done, and will always do: They are cutting and running. This is called *innovation*.

In 1995, blaming everything from a proposed federal forest plan to high demand to disease to forest fires, Boise Cascade said it was seeking timber outside the Northwest. Boise Cascade CEO George Harad admitted that private timber was "being cut at an unsustainable rate," and corporate spokesman Doug Bartels admitted that "in any [public timber] scenario we're still going to have to look offshore for timber supplies."[39] Boise Cascade vice president Dick Parrish and other corporate officials had toured Siberian timberland. They were also considering radiata pine forests in South America, South Africa, and the South Pacific (principally New Zealand). Once again, let's translate: despite all rhetoric about sustainable forestry and community commitment, the real commitment is to deforestation, to cutting and running.

By encouraging globalization, free trade agreements encourage the destruction of human and nonhuman communities.

They do this by reducing tariffs and removing worker, consumer, and environmental protections, all in order to encourage multinational corporate investment in foreign countries. "Barriers to trade" that violate World Trade Organization rules include, among many others: export controls used to protect local jobs or native forests (meaning no region can demand local forests be used to benefit local people); requirements for the reuse and recycling of paper, or for controlling the use of packaging materials (meaning no region can require recycling); certification and labeling of forest products (meaning no region can require forestry to be even *remotely* sustainable); performance requirements for foreign investors such as requiring investors to take local partners, to hire local workers, to make certain levels of investment, or to transfer environmentally beneficial technology to the local government or to local companies; provisions allowing discrimination against foreign investors based on poor environmental records (meaning even if Weyerhaeuser has clearcut more than 4 million acres just in the United States, which by its own admission it has,[40] no nation can use that information to effectively protect itself and its forests); restrictions on foreign ownership of forest land; and even measures intended to control the invasion of exotic species. All of these are considered violations of "free trade" rules, and nations that enact them are subject to economic sanctions.[41]

Free trade increases consumption.

According to studies by the World Trade Organization's Committee on Trade and Environment and the American Forest & Paper Association, tariff elimination for forest products could generate a 3 to 6 percent growth in consumption and an annual increase of $350 million to $472 million in worldwide trade for selected forest products in key markets. They consider this good news. I'm not so sure forests—or normal humans—would agree.[42]

Free trade and globalization lead to more subsidies.

Subsidies to the wood products industry include publicly funded timber-sale administration, road construction, and forest fire and erosion control; exemptions from property taxes; tax breaks for job-destroying log exports; retraining for laid-off mill workers; and hazardous waste cleanup at abandoned mill sites. Increased corporate mobility inevitably leads to increased job blackmail. Job blackmail occurs when corporations threaten to close or move an operation unless workers accept wage reductions and local governments offer greater subsidies. These local governments often compete with each other to attract or retain timber corporations by offering reductions in property, sales, or other taxes, free sites for mills and port facilities, and low-cost loans or grants for equipment and modernization. For example, Ohio and Kentucky have been engaged for years in a multimillion-dollar competition to subsidize International Paper's "restructuring." The public pays for deforestation and the shifting of jobs from one place to another, and the taxpayer and worker always lose. Wouldn't it be nice if normal people had this same ready access to free housing as corporations have to free mill sites, and to free food as corporations have to subsidized raw materials?

As free trade agreements encourage the closing of domestic mills, a new layer of subsidies to globalization has been created: The U.S. Department of Labor gives "NAFTA aid" to laid-off workers for job retraining, relocation reimbursements, and extended unemployment benefits when plant closures have been caused by cheap imports or the relocation of operations to Canada or Mexico.[43]

Meanwhile the World Bank and multilateral development banks, flush with endless taxpayer monies, provide loans for destructive and unnecessary timber operations. When local people fight back against the invasion of foreign corporations, the U.S. State Department's Overseas Private Investment Corporation provides government-backed political risk insurance, such as for International Paper's takeover of a paper mill in Russia.[44]

The game is rigged.

But it's not a game. Or rather it's only a game for the "winners." For the "losers" it's poverty and degradation, life and death.

A voice from Papua New Guinea: "You white people use sawn timber to build your houses. We Niugini use black palm for flooring. We use cane instead of nails. We use Kunai to make our roof instead of iron. Machines of the company have spoiled our black palm trees, our cane, and the dozers have trampled our Kunai land. Gone is the Malou we use to make our traditional clothes for sharing our customs with the other villages. Machines have spoiled our land and our tradition. Money is no compensation."[45]

Transnational corporations are eating the world. One of their strengths as a tool of the rich is that they kill so effectively at great distances. Those who profit never have to see the devastation they cause. Another of their strengths is that because they do not actually exist, because they are legal fictions, they can never be killed. Worse, because they have no bodies—being instead the "embodiment" of greed, and because they, like the rest of our culture, are an attempt to separate themselves—for whatever stupid, insane, murderous, and suicidal reason—from the world that surrounds them, they effectively never need to stop growing.

So corporations grow huge as forests grow small. Our lives and local economies grow ever more controlled by these ever larger, ever more distant, and ever more imperious fictional entities.

And they are eating the world.
It's all very strange, very sad, very stupid.

Transnational corporations and those who run them have no allegiance to place. Corporations based in the United States deforest Southeast Asia. Corporations based in Southeast Asia deforest South America. Corporations based in Europe deforest Africa. It doesn't stop.

Our problem with globalization is not based on abstract economic and political theory. We are talking about actual companies and real forests. The largest "landowners" in the world are timber companies.

The Canadian corporation Abitibi-Price owns a milllion acres in the United States and Canada, and holds cutting rights to 19 million acres more. Barito Pacific holds 2 million of Indonesia's 21 million acres of forestry concessions.[46] Canadian Pacific Forest Products owns or holds tenures on 24 million acres. The Japanese paper manufacturer Daishowa controls nearly 10 million acres of timberland in Alberta and British Columbia, Canada.[47] Karl Danzer of Germany controls more than 7 million acres worldwide, including concessions accounting for 10 percent of Zaire's forests and 40 percent of its logging, and an Ivory Coast concession of almost a million acres. In the 1980s Danzer also operated in Brazil and Argentina.[48] From its concession of 750,000 acres in Gabon's Bee forest, the German corporation Glunz obtained okoume logs for its Isoroy subsidiary, the largest tropical plywood producer in Europe. Seeking more "raw material," Glunz also gets logs from Cameroon, Central African Republic, Equatorial Guinea, and the Congo. Hyundai signed a thirty-year logging agreement in the Primorskiy Krai in Siberia in 1990; much of the timber is exported as raw logs; Hyundai has also expanded into the upper Bikin watershed.[49] International Paper has controlled as much as 12 million acres across the United States. The Timber Group branch of

John Hancock Insurance Company controls more than 3 million acres in the United States and Australia.[50] The Japanese conglomerate Mitsubishi controls more than 20 million acres of timberland in Australia (wood chips), Brazil (plywood), Canada (pulp and paper and chopsticks), Chile (wood chips), Papua New Guinea, and the United States (wood chips). New Oji Paper, associated with Mitsui of Japan, controls 17 million acres of timberland and operates wood chip, pulp, and paper mills in Canada, Germany, New Zealand, the United States, Australia, New Zealand, Fiji, Papua New Guinea, Vietnam, Brazil, Thailand, Chile, and Indonesia. The Malaysian timber company Rimbunan Hijau owns or controls more than 9 million acres in Papua New Guinea (and accounts for more than half of PNG's total timber cut), Malaysia (2 million acres), Brazil (more than a million acres), Khabarovsk Russia (almost a million acres), Cameroon, and New Zealand. Rougier Ocean (a French corporation) controls more than 300,000 acres and a sawmill and plywood plant in Cameroon, and exports logs to France, Italy, Spain, and Japan. Half of Cameroon's debt to France was cancelled in 1994 in exchange for Rougier Ocean's access to Cameroon timber, while French President Mitterand's son was, coincidentally enough, a shareholder in Rougier's Cameroonian subsidiary. Rougier also has the largest foreign holding in Gabon timber (more than 90 percent of Gabon's timber cut is exported as raw logs). The Malaysian company Samling signed a 1994 agreement giving it access to almost 5 percent of Cambodia's land area and 12 percent of its remaining forest. Samling has entered into multimillion-acre agreements in Brazil. In 1991 Samling and the Korean company Sung Kyong obtained a twenty-five-year license to log 4 million acres in Guyana. Samling also operates at "home" in Malaysia, with rights to cut more than 3 million acres. The Malaysian corporation Tenaga Khemas has a concession of almost 2 million acres in Guyana.[51]

And so on.

• • • •

A voice from Malaysia, a statement signed by sixty-one tribal Dayak leaders: "Some people say we are against 'development' if we do not agree to move out of our land and forest. This completely misrepresents our position. Development does not mean stealing our land and forest. . . . This is not development but theft of our land, our rights and our cultural identity."[52]

What we said earlier, about how corporations never need to stop growing, is not precisely true. They will not stop growing until they have consumed the world. Surely they will stop then. And so will the forests. And so will we.

The Failure of Solutions

> The whole fabric of society will go to wrack if we really lay hands of reform on our rotten institutions. From top to bottom the whole system is a fraud, all of us know it, laborers and capitalists alike, and all of us are consenting parties to it.[1]
>
> *Henry Adams, 1910*

We do not know the true state of the world's forests. And even if we did, what good would it do? The most recent forest inventories use the latest satellite imagery, but when half the logging is illegal, what use is an inventory?

Global consumption of wood is up 50 percent since 1961.[2] The foresters at the United Nations predicted that wood consumption would rise 23 percent between 1996 and 2010, and paper consumption would rise 30 percent.[3] Like a smoker cutting down on New Year's day, any slowdown in the consumption of wood and paper due to economic recessions will soon be lost in the next compulsive frenzy.

Japan is responsible for more than 40 percent of the world's tropical wood trade. A third of the tropical wood that Japan imports is plywood used once or twice as cement forms and then thrown away.[4] Forty percent of Japan's plywood supply, which comes from Indonesia, Malaysia, and Russia, is of illegal origins.[5] It is painfully obvious that the Japanese do not have to use tropical plywood as disposable cement forms.

More than three-fourths of the tropical timber used in the United States is in the form of lauan plywood. It's used for doors, under floors, and as furniture backing, signs, and movie and theater sets. And the United States is the world's biggest importer of mahogany. It is painfully obvious that New York City simply does not have to use Brazilian ipe for its park benches or to deck its municipal boardwalks. Guyanan greenheart does not have to be used as pilings in the city's harbors and marinas. African purpleheart does not have to be used for the crossties under New York's subway tracks. Mahogany smuggled out of the Amazon and teak logged with Burmese slave labor do not need to be used for conference tables and desks.[6]

There seems to be no limit to the amount of paper our culture can consume. We mentioned before that annual world paper consumption increased from 15 million tons in 1910 to 463 million tons in 1996. To consume that 463 million tons requires cutting 2 billion cubic yards of wood covering 2 million acres of forestland.[7]

Paper consumption grows far faster than population. And the consumption is not spread evenly. The average person in America consumes almost 700 pounds of paper per year; the average in Great Britain and Japan is 330 pounds per year; the average in the nonindustrialized world is 12 pounds per year.[8]

Not enough paper is recycled. Less than half of U.S. paper was recovered for reuse in 1997, and recovered waste paper accounted for only a third of the U.S. industry's fiber needs. A third of U.S. printing and writing paper was recovered in 1997, most of it used to produce tissue and paperboard. Three-quarters of the corrugated cardboard boxes were recovered, and two-thirds of the newspapers. Even U.S. government agencies, which could jump-start the recovery of wastepaper in the United States, are required to buy no more than 10 percent recycled paper.[9]

But here's the kicker: two-thirds of the world's paper is made into packaging, tissue, and other disposable products.[10] We are committing genocide, ecocide, and global suicide by wiping our behinds with ancient forests.

• • • •

An Iban woman from Sarawak, after a longhouse was resettled to make way for a dam, said of her people: "I believe they will cry forever because they have lost their lovely land."[11]

Pulp and paper manufacturing has expensive and deadly side effects. It consumes huge amounts of water and energy. It produces toxic chemicals that pollute air and water and soil. It is a major contributor to mountains of solid waste. With all these problems, the paper industry has the lowest levels of research and development of any major manufacturing sector in the United States.[12] Less than one percent of U.S. paper fiber comes from sources other than wood. Worldwide, only 10 percent of virgin pulp and 6 percent of all paper fiber is not from wood. Yet China manages to get 60 percent of its paper fiber from alternative sources. Paper fiber historically was and still can easily be made from agricultural residue like wheat, barley, and rice straw, corn and sorghum stalks, and sugar cane bagasse. Germany and other European countries have lifted their bans on hemp fiber, and more than 800,000 acres of hemp fiber are being cultivated in the northern hemisphere.[13] All of this helps slow the acceleration of deforestation.

The kraft mills that use chlorine are expensive to build, and it's expensive—tens of millions of dollars—to upgrade a kraft pulp mill's machinery and industrial processes. But the changes are worthwhile. Besides reducing pollution, the alternatives to chlorine bleaching also reduce energy use and other operating costs, and produce higher yields of pulp. The European pulp industry made necessary changes years before the U.S. industry. The U.S. EPA and U.S. paper industry colluded to suppress the dangers of dioxins and the paper industry's role in exposing the public and environment to it. One judge declared that they had collaborated to "suppress, modify, or delay the results of the joint EPA/industry [dioxin] study or the manner in which they are publicly presented."[14]

The U.S. government and industry have delayed the inevitable changes by focusing on the measurement and control of specific chemicals, rather than on the more productive and sensible prevention of whole classes of toxic chemicals.[15] The obvious (and intentional) limitations of this approach are stunning. Wood-based panels such as oriented strand board, medium density fiberboard, particleboard, and plywood are manufactured with cancer-causing nerve poisons such as phenols and formaldehyde. Lumber is preserved with poisons such as arsenic and pentachlorophenol. Many of these toxins are regulated. That may sound reassuring, but this is how regulation generally works: A regulatory agency without enforcement resources or political will issues legal notices containing guidelines on how the toxin may be used, and how it may be disposed of. That's it. I hope you enjoy your cancer. Sometimes corporations are fined, but the fines are small enough that many corporations, such as Weyerhaeuser, treat them as routine costs of business, much like advertising. It is far cheaper to pay fines and continue to routinely poison human and nonhuman communities than to fix the problem. So the problem doesn't get fixed.

Of course.

Numerous ways to rate companies on their pollution performance have been devised. The EPA requires manufacturing facilities to disclose how many pounds of toxic chemicals they release. Public interest organizations such as the Council on Economic Priorities have published profiles of paper manufacturers that calculate how many pounds of toxic chemical a company emits per employee or per dollar of revenue.

Big deal. People continue to get poisoned. Waterways continue to get poisoned. Landscapes continue to get poisoned.

Deforestation accounts for about half of the human releases of carbon dioxide, one of the major causes of global warming.[16] The

American Forest & Paper Association, the International Council of Forest and Paper Associations, the World Business Council for Sustainable Development, and the World Resources Institute have developed a method for calculating greenhouse gas emissions from pulp and paper mills.[17]

The gaining of information is a good thing. But the world continues to burn. The problem is not a lack of information, but one of having ignored the deeper knowledge of how to live a different way.

The World Resources Institute (WRI) recently came up with a method for integrating environmental liabilities into financial performance accounts. For example, the measurement would include as financial liabilities all of a company's foreseeable costs for complying with pollution regulations. Once you hear about it, it seems so obvious you can't believe this hadn't already been done and been widely accepted. As the authors point out, it doesn't matter what you think about the environment; it's simply prudent business to consider all your liabilities. Yet financial analysts and investors have focused routinely (and exclusively) on extracting profits as quickly as possible. Shareholders aren't liable for their company's liabilities, and directors and managers claim a "fiduciary" duty to deliver maximum dividends to shareholders. And WRI claims that investors have been unable to include environmental risks and opportunities in their decision-making "largely because of the lack of a feasible methodology for translating environmental performance into financial terms."[18]

This statement, which isn't so much false as myopic and forgiving, perpetuates the focus on technical (in this case, accounting) methods of regaining simple sanity. New accounting methods are long overdue. Companies complain about "command-and-control" environmental regulations, yet refuse to adopt sound business practices, much less sound environmental practices. It takes an environmental think tank to show multi-billion dollar paper companies how to integrate their liabilities into their financial accounts.

This is all insane.

Can we accept another, more straightforward approach? No poison should be made for which there is no antidote. No industrial process should occur which fouls our water supplies. No one should kill workers or neighbors or activists or peasants or the indigenous or the landscape in the process of manufacturing park benches, Kleenex, or grocery bags. Why are these commonsensical and ethically self-evident principles so impossible for so many to practice?

The problem is not and has never been a lack of accounting methodologies or industrial know-how; the problems are denial, recalcitrance, and apathy. The solution isn't technical, but political. The solution isn't even political but social. The solution isn't even social but psychological. The solution isn't even psychological but perceptual. The solution isn't even perceptual but spiritual. The problem is our entire way of living and relating to the world.

Nonetheless, let's try on this political solution: Arm EPA regulators with the best methodologies, and force the U.S. Securities and Exchange Commission and the Department of Justice to sue and bankrupt any company (including its shareholders, directors, and managers, who might also be jailed) that cheats its investors, the taxpayers, and/or the human and nonhuman community by ignoring human rights and the environment.

What a concept.

Canadian and American taxpayers pay for destruction of their forests. Brazilian and Guatemalan peasants suffer the depradations of their landlords. Future generations will pay for their ancestors' gluttonous consumption. Lemurs and tigers pay through their persecution—and extirpation—by humans.

Some of these externalizations of the worldwide timber

industry are unintentional, and cause the destruction of forests as an unplanned consequence. Others are deliberate, and are intended to boost or maintain profits of various wood and paper industries by externalizing costs or increasing consumption of wood and paper products to match the industries' overcapacity, making human "demand" keep up with a supply created by megamills.

From 1996 to 1998, the U.S. wood and paper products industry took $3.6 billion in profits. They paid $500 million in taxes and received $759 million in tax breaks. Two dozen wood and paper companies paid less than nothing in taxes; they actually received taxpayer money. For example, in 1998, Weyerhaeuser's taxes were a *minus* $9.5 million.[19] In 1994 the state of Ohio granted a package of subsidies to International Paper, including a $7 million loan and a $700,000 grant to buy equipment, a $3.4 million tax credit, $420,000 in sales tax exemptions, and a $90,000 job-training grant.[20]

The U.S. State Department's Overseas Private Investment Corporation and the U.S. Export-Import Bank fund deforestation overseas, from Russia to Indonesia to Chile. Japan subsidizes the destruction of forests and their replacement by sterile plantations around the world—in Russia, Okinawa, Indonesia, Brazil, Australia, and Vietnam. European nations fund roads and dams that chip away at what's left of the African rainforests.

Planting trees and environmental monitoring of timber corporations are nearly meaningless when tax monies are used to build logging roads and log export docks, to insure politically risky logging operations in regions torn by civil wars, to exempt paper corporations from environmental regulations, to loan tractors and pulping machines to communities which have managed so far to protect their forests, and to provide police and military to attack or kill those who resist.

These subsidies must be changed. Imagine if the same amount

of money was thrown at actually trying to help forests—not through bogus schemes where those in power change the names of destructive actions to make them more palatable (remember *temporary meadows?*) but through programs that serve the forests and all who live there?

What a concept.

Here are the words of Along Sega, Penan headman from Long Adang, Sarawak, rejecting money from a Limbang Trading Company manager, after the graves of Along Sega's family were destroyed: "I told him, even if I have to die of any cause I shall not trade the bodies and souls of my parents and relatives to save mine because our bodies, dead or alive, are not for sale. I refused the money and pleaded with him also that if you have so much money already please don't come here to take our land. But he just shook his head, laughed and replied, 'We have been licensed to work on this land. There is no such thing as your land in the forest because forest belongs only to the government. Take this money or you get nothing.' I still rejected the money."[21]

Forestry should protect forests. The idea is simple, the rationale obvious. Forestry should be sustainable. This is not a technical problem. Timber cutting should not destroy forest "products" such as mushrooms, clean water, and oxygen. Timber cutting should not be so rapid as to destroy the capacity of a forest to produce fiber. (There are no productive third-growth forests.) Every proposal to cut timber should include an ecosystem-wide environmental impact assessment—a real one, not more "no significant impact" bullshit—as well as a detailed and honest evaluation of how the proposal does or does not compromise the forest's fiber or other productive capacities, and more importantly what it means to the overall health of the forest. The operation should be guided by forest-specific criteria of sustainability and conservation. And the proposal should be certified as accurate by an independent party who has no stake in the timber operation's profitability.

In fact, millions of acres of commercial timberland are today being certified as sustainably managed. Wholesale purchasers and retail customers are beginning to be provided with a clear chain of custody, from the forest to the mill to the point of sale. More than 30 million acres have been certified for sustainable management, and 50 million acres are covered by a "formal, nationally approved forest management plan covering a period of at least five years." Unfortunately, 50 million acres is only 6 percent of the world's forest area, and of course many of those forest plans are never implemented. Ninety percent of these certified forests are located in the industrialized nations of the United States, Finland, Sweden, Norway, Canada, Germany, and Poland, leaving southern forests and forest-dwellers unprotected.[22]

Two of the major certification programs are the Forest Stewardship Council (FSC) and the Pan European Forest Certification Council (PEFCC). The FSC was established by environmental groups and timber companies to show consumers which wood products are from "environmentally acceptable, socially beneficial, and economically sustainable" sources.[23] The FSC is developing principles and ecosystem-specific criteria of good forest management. But the FSC receives money from the companies it certifies. The result? A recent investigation of FSC's programs in Brazil, Canada, Indonesia, Ireland, Malaysia, and Thailand found that FSC had certified companies that are "implicated in gross abuses of human rights, including the torturing and shooting of local people; are logging in pristine tropical rainforest containing some of the world's most endangered wildlife species, such as the Sumatran tiger; and have falsely claimed to comply with the FSC's audit requirements, such as by allowing uncertified wood to be labeled with the FSC seal of approval."[24]

The Pan European Forest Certification Council was established by European forest owners and industry representatives. Ninety-eight percent of Finland's and two-thirds of Norway's forests have been certified sustainable, which proves little more than that an organization exists to certify sterile tree plantations.[25]

When Boise Cascade, Weyerhaeuser, Rimbunan Hijau, Samling, and other notorious corporations are certified, does certification mean anything? Do the chains of custody papers mention slave labor, smuggled logs, or toxic pollution? ForestEthics and the Rainforest Action Network should be (and are) hounding the smugglers and butchers to the ends of the earth, and shaming lumber stores into offering certified wood. Yet so long as we maintain our current system of cultural rewards, there will always be politicians willing to give public resources away, loggers willing to cut ancient forests, consultants willing to falsify chain of custody papers, and consumers willing to buy the cheapest product, regardless of who was killed to get it. And in the end, only low-tech forestry operations for local consumption will ever be truly sustainable.

In the meantime, it might not be a bad idea to require wood products operations to provide auditors and consumers with a clear and complete chain of custody. Nor would it be a bad idea to expand the protection of forest resources and phase in the return of natural forests.

What a concept.

Don't hold your breath.

We need to protect ecosystems—for their own sake most of all, but also for our own. More than ten thousand plant species are used as traditional medicines; most of them are gathered from the wild, from the forest. Eighty percent of the people in nonindustrialized nations depend upon them for their primary health care; 25 percent of modern drugs are also derived from these plants.[26]

Restoration forestry restores ecological and genetic diversity and soil nutrients, structure, and biology. It can restore abused and fallow forest and agricultural lands. We can close roads to reduce erosion and prevent further entry by vehicles and other machinery. We know how to thin dog-hair stands to restore natural forest structures. We know enough to understand the need to control exotic pests and reintroduce the full natural diversity of trees and shrubs and fungi and animals and human forest-dwellers. We can restore streams to reduce erosion, restore anadromous fish, and so on. We know how to do all of this. So do forests.

Here is a definition of restoration forestry: "Restoration forestry assists nature to heal degraded forests and bring them back to a state of biological productivity, biodiversity, ecological stability and resilience. Restoration forestry means increasing the area under forest cover and increasing the age classes, the standing volume and the diversity of forest ecosystems. It means careful harvesting [sic] methods that minimise disturbance of soil and plant communities. It means that many more people will have to be employed in the woods, not less; using smaller machines and more reliance on draft animals. It means smaller mills and more value-added processing close to the wood source. It means minimal waste, maximum recycling, and the development of non-tree paper pulp and alternative building materials. It means more people caring for the forest and researching its complex processes, so that we can ever refine our management/dance with the forest. Restoration forestry leads to a steady yield of high value timber. Clearcutting and/or short-rotation forestry leads to a periodic return of low-quality timber. Restoration forestry makes much better ecological sense and it makes better economic sense."[27]

We need to distinguish restoration forestry from restoration ecology. Forestry is for producing a supply of wood. If you are an intelligent forester, you would restore tree stands (such as plantations) to a natural, optimal fiber-producing capacity. But you are still a forester, looking for wood fiber.

An ecologist would protect or restore fully-functioning forest ecosystems, and consider fiber production for human use to be completely subordinated to the full range of natural ecosystem functions.

Once again, what a concept.

We must move away from industrial forestry and toward restoration forestry. We must then move away from restoration forestry and toward restoration ecology.

We must also move away from globalization, toward community forestry (hell, toward *all* forms of community).

"Community forestry is a village-level forestry activity, decided on collectively and implemented on communal land, where local populations participate in the planning, establishing, managing and harvesting of forest crops, and so receive a major proportion of the socio-economic and ecological benefits from the forest."

"Successful community forestry requires . . . genuine popular participation in decision-making. . . . Experience has proven time and again that participation is more than a development cliché; it is an absolute necessity if goals are to be met. But working with people rather than policing them is a new role for many foresters."

"Community forestry has the following characteristics: The local community controls a clearly and legally defined area of forest; the local community is free from governmental and other outside pressure concerning the utilisation of that forest; if the forestry involves commercial sale of timber or other products, then the community is free from economic exploitation of markets or other pressure from outside forces; the community has long-term security of tenure over the forest and sees its future as being tied to the forest."

"Community forestry, social forestry and rural development

forestry are more or less equivalent and reflect Abraham Lincoln's view of democracy—government of the people, by the people, for the people."

"The political dimension of community forestry makes it a venue for people's struggle against domination and exploitation of the community's resources by 'outsiders.' Ecology, equity and social justice are part of this struggle."[28]

Of course.

There have been many international treaties and agreements, systems of trade rules, and declarations of intent to promote sustainable economies and to protect forests, endangered species, and the human rights of forest-dwellers. Some of these agreements and treaties include: the 1994 Commission on Sustainable Development & The World Summit for Social Development;[29] the "Rio Forest Principles" drafted at the 1992 United Nations Conference on Environment and Development (UNCED), officially entitled *Non-Legally Binding Authoritative Statement on Principles for Global Consensus on the Management, Conservation, and Sustainable Development of All Types of Forest*;[30] the 1992 Convention on Biological Diversity;[31] the Convention on Conservation of Migratory Species;[32] the Convention on International Trade in Endangered Species (CITES);[33] the International Tropical Timber Agreement;[34] the Ramsar and World Heritage Conventions; the UNESCO-MAB Biosphere Reserves Programme; and so on.[35]

Really, the only words you need to read in the preceding paragraph, and the only words you need to read in most of these treaties, are the beginning words of the official title of the Rio Forest Principles: *Non-Legally Binding*. That tells you almost everything you need to know.

While there has been great passion, ingenuity, and effort put into drafting and reaching consensus on these agreements, together they form a hopeless mass of contradictory statements and intentions, combined with lofty abstractions. For example, the Rio Forest Principles include these statements:

"National policies and strategies should provide a framework for increased efforts, including the development and strengthening of institutions and programmes for the management, conservation and sustainable development of forests and forest lands."

"Efforts should be made to promote a supportive international economic climate conducive to sustained and environmentally sound development of forests in all countries, which include, interalia, the promotion of sustainable patterns of production and consumption, the eradication of poverty and the promotion of food security."

"Efforts should be undertaken towards the greening of the world."

"New and additional financial resources should be provided to developing countries to enable them to sustainably manage, conserve and develop their forest resources."

The most important word, repeated in every statement here, is *should*. Not *will*. Not *must*. *Should*. As usual, we get words while those in power get the forests.

The official statements are a greenwashing of the sustainability façade that covers the global industrial trade in wood and paper. The agreements on biodiversity and forest protection are completely trumped by binding treaties and trade rules coming out of the General Agreement on Tariffs and Trade (GATT) and its successor, the World Trade Organization (WTO).

But even if the forest conservation agreements were made consistent—and that's not likely to happen—and rewritten to actually address the problems head-on, there would have to be a fundamental shift in political will, government funding, and enforcement to back up the fine rhetoric.

Even if these agreements and rules were implemented, what

we would see are minor technical, economic, and legal reforms rather than genuine solutions to the forest crisis. Eventually the decisions about land and resource use need to be controlled by local peoples who know, love, and depend on the forests.

We pride ourselves on our democratic freedoms, but even our façade of democracy stops at the border. Half the extant human languages are in Papua New Guinea, the Congo Basin, and the Amazon.[36] Democracy would look quite different if those who spoke languages other than English were allowed voice. Heck, democracy would look different if those who spoke languages other than high finance were allowed voice.

It is fashionable to condemn "banana republics" for having unequal distributions of land, but U.S. land distribution is no better. In Brazil, 2 percent of the farms occupy 54 percent of the land.[37] In Honduras, two-thirds of the arable land is owned by 5 percent of the farm owners.[38] In the United States, the top 5 percent of *landowners* (not 5 percent of total population) own 75 percent of the land,[39] worse than the ratio in Honduras. In California in 1970, 58 percent of the farmland was owned by twenty-five owners (*not* 25 percent).[40] And as we've seen, many of the largest landowners in the United States and around the world are timber corporations.

The landless worldwide are surviving on a fraction of the acreage that was once theirs, acreage that is still available, being held to keep people landless—so they will be cheap laborers—and for speculative real estate purposes.

We don't need "public participation, consensus, and collaboration," or "community forestry" programs run by corporate and government elites; we need local control of land and markets. Adam Smith's invisible hand of the market only worked when the market was local, face to face, voluntary, transparent, low-tech, and based on ethical, mutual relationships. It's been a long time since that was the case.

. . . .

Grassroots organizations and alliances have drafted their own statements regarding forest conservation. Many of them are clear in stating that human rights and environmental protection come before economic "development." The Declaration of the World Rainforest Movement in Penang, Malaysia in April 1989, for example, called upon the United Nations and national governments "to empower forest peoples and those who depend upon the forests for their livelihood with the responsibility for safeguarding the forests." It asked the UN to give them "a decisive voice in formulating policies for their areas," to "reject social and economic policies based on the assumed cultural superiority of non-forest peoples," and "to halt all those practices and projects which would contribute either directly or indirectly to further forest loss."[41]

Ten years later, in June 1998, the Montevideo [Uruguay] Declaration pledged "support for an international campaign to support local peoples' rights and struggles against the invasion of their lands by these plantations; encourage awareness of the negative social and environmental impacts of large-scale industrial monocrop tree plantations; and change the conditions which make such plantations possible."[42]

More recently, a declaration signed by many groups at the World Summit on Sustainable Development in Johannesburg, South Africa in August 2002 called for "recognition of indigenous and other forest peoples' territorial rights; agricultural land reform; a moratorium on external debt repayment; a legal international instrument for corporate control; equitable North-South trade relations; decreasing overconsumption in the North; profound reform of multilateral institutions (IMF, World Bank, regional banks) to put them at the service of people and the environment; a moratorium on oil, gas and mining activities in tropical forest areas."[43]

Written by ecologists, community leaders, and indigenous peoples, these declarations are a great improvement over the treaties signed by nation-state governments. They are wise and heartfelt calls for a return to sanity and sustainability. Sadly, it is difficult to imagine them being enforced, seriously considered, or even allowed by the kinds of governments now in power.

Rejecting Gilgamesh

> What we are doing to the forests of the world is but a mirror reflection of what we are doing to ourselves and to one another.
>
> *Mahatma Gandhi*

We have some declarations of our own. Immediately leave remaining frontier forests alone, and confine industrial forestry to existing plantations. Soon, once we have learned how, restore most and then all of the plantations to natural forests. This work could be done by restoration ecologists, who, like traditional forest-dwellers, are grounded in their specific local natural communities. Restoration ecology will be one step toward recovering indigenous knowledge and techniques, which is always specific to place. The purpose of restoration is not fiber production, even sustainable fiber production, but restoring ecosystems and their humans to their natural local patterns and processes.

Perhaps most important of all, relinquish control of land to those who belong to the land. Satellite data has shown that where indigenous people hold land title, there has been less forest destruction. Give back the land to the humans and nonhumans who live there, and who have lived there for a very long time, who belong to the land. Give it back to those from whom those in power stole it, and from whom they continue to steal it.

We can already hear the obvious objection: "That's absurd. It's not practical. Be real. Colonization has turned forest-dwellers into landless peasants working agricultural plantations, and industrialism has turned them into an urban working class for factories. This may have been regrettable, may have been unjust, but it happened. It's done (well, not exactly, since it's an ongoing process). Get over it. We've committed ourselves to industrial civilization, and we've got to see it through to the end. Do you think you can put trees back on stumps, unroll roads from forests, let peasants and factory workers become forest-dwellers again? What you are proposing is unrealistic, and in fact dangerous."

It is always odd to hear words like *realistic* and *dangerous* coming from the mouths of those who value money over life, who say things like "we must balance the needs of our economic system versus the environment" (which of course is a tacit and entirely accidental acknowledgment of what we all know: the needs of the economic system are in direct opposition to the needs of life).

What is real?

What do you love?

What do you fear?

What do you need?

The problems we face look far different, depending on who you are, on what you love, what you fear, what you need.

What are your actual and self-perceived relationships to forests, to power, to society?

Weyerhaeuser shareholders presumably perceive it in their best interest to maximize dividends by liquidating current assets: in other words, cut to the last stick. Shareholders (at least as shareholders) do not relate to their actual assets (trees), much less to forests. They relate not to forests, nor trees, nor even to productive goods and services, but to wealth as dividends.

Most of us environmentalists—grassroots environmentalists, not the corporate environmentalists of the Sierra Club and Audubon national offices—are holding on by our fingernails, trying to save whatever scraps of forests we can, using whatever tools we can cobble together, putting our hearts and minds and time and sometimes bodies between the chainsaws and the forests we love. And we are praying, every moment of every day, for civilization to end. For this culture to run out of oil, to collapse in on itself. For this long and awful nightmare of deforestation and dispossession to end.

Some social justice activists see inequality as the root of the problem, and believe that if we just bring sufficient and equitable "development" to the poor, then the world's problems may be solved. In their hearts some hold hope for a great proletariat revolution that will bring justice to all. Does their "all" include those humans and nonhumans who live in forests?

What are the hopes and fears of Southern elites trapped between dispossessed countrymen and gringo bankers? Where do forests fit into their dreams, their nightmares? What do they want for and from these ancient trees? Or do they think of them at all?

And what is the perspective of the slum-dwellers in Brazil and other countries, dispossessed by five hundred years of colonial and neoliberal conquistadors? What do they want? What do they need? What do they fear? What can we—those of us who have been privileged enough to learn how to read books like this, printed on the flesh of trees torn from the soil where these people once lived—what can we do to help?

And what of the forest dwellers themselves? What of the indigenous? What do they want? What do they need? What are their relationships to the forests where they live? Can we help them, if only by leaving them to their own good and sufficient lives?

Tigers. Sumatran Rhinos. Orangutans. Hazel's forest frogs. What do they need? What do they want? How can we help?

And the trees. Redwoods. Lodgepole pines. Port Orford Cedars. American Chestnut. Lauan. Mahogany. Ipe. Greenheart. Purpleheart. Teak. What do they want? What do they need? What is your relationship with the trees and forests where you live?

• • • •

It is our present course that is unrealistic, and doomed to a nightmare failure so complete that perhaps only those who are forced to live these horrors—the humans and nonhumans of the forests, and the forests themselves—are able to even partially comprehend it.

Plato observed, back when the soil of his home still bore recent memory of lions and forests and the people of the forests, "What now remains compared with what then existed is like the skeleton of a sick man, all the fat and soft earth having wasted away, and only the bare framework of the land being left. . . . There are some mountains which now have nothing but food for bees, but they had trees not very long ago. . . . There were many lofty trees of cultivated species and . . . boundless pasturage for flocks. Moreover, it was enriched by the yearly rains from Zeus, which were not lost to it, as now, by flowing from the bare land into the sea; but the soil it had was deep, and therein it received the water, storing it up in the retentive loamy soil, and . . . provided all the various districts with abundant supplies of spring-waters and streams, whereof the shrines still remain even now, at the spots where the fountains formerly existed."[1]

Deforestation boils down to power. Those who deforest do so because they are supported by the full might of the state. It is ludicrous for anyone to suggest that those who stole these lands by force, and who maintain control by force, and who deforest at the point of a gun, will give the land back to its rightful human and nonhuman owners because it is the right thing to do, the sane thing to do, the human thing to do, the nonsuicidal thing to do. No.

They will not leave the forests, and leave the forests alone, until either the forests are gone, or until those of us who love the land force them out of the forests.

We do not hope to stop deforestation with this book alone, any more than we can stop it by filing timber sales, compiling corporate profiles, voting, doing tree sits, or sending money to good organizations like Amazon Watch or Cultural Survival.

We do not know how to stop deforestation. We do not know how to get deforesters out of the forests. No one else—forest-dwellers or civilized—has figured that out either, or surely the deforesters would have been removed by now.

But we do know this. Once people see deforestation for the atrocity that it is, they will then stop those who continue to destroy. It is for this we wrote the book. It is to this we have dedicated our lives.

What is your relationship with the future?

There is nothing humans can do to maintain industrial wood and paper production, and maintain forests too. The crisis will resolve itself when civilized humans walk away from doing what they have been doing.

Those in power won't stop deliberately. Never forget Red Cloud's warning: "They made us many promises, more than I can remember. But they never kept but one. They promised to take our land and they took it."[2]

What, then, can you do if you are of good heart? You can fight to keep this particular tree standing, this particular forest functioning. You can help open your friends' and coworkers' eyes to the wonder and intrinsic value and legitimate standing of forests and forest-dwellers.

We don't need to stop the forest crisis. Nature will stop it. As the global economy becomes more chaotic and the societies addicted to it become more impoverished, the best we can do is to keep some doors open, prevent those in power from causing those we love to go extinct. Defend what is important, undermine what is superfluous, and destroy what is destructive.

We can consume less. We can eat less meat, drink less coffee.

We can eat locally grown foods. We can make our own food, clothing, and shelter, so as not to deny others the right and ability to provide for themselves. Those of us in the heart of empire can work to undermine the social and political basis for our inordinate power over others. We can work to implement radical equality.

We can spend time in forests. We can ask the trees—and forests—what they want. Early on, this book described a walk through an old-growth forest, and it ends on the same note, by inviting you to do the same. We humans came from forests, and to them we will return.

We have been the obedient servants of Gilgamesh for five thousand years. We have cut a path of destruction, ignored the spreading deserts, disregarded the disappearing animals, the fouled air and water, the warming planet. We have destroyed most of the earth's natural forest cover, and we pretend we can live without it. The story we have been handed says that Gilgamesh defeated the forest protectors and that the forces of civilization won the battle for the forest, but it's not true. The epic is not over, and Enlil's curse will not be lifted until we reject the easy and false promises of Gilgamesh, and return with respect and humility to the forests.

ACKNOWLEDGMENTS

George Draffan:
Thanks to the grassroots clients of the Public Information Network for their frontline work on behalf of the world's forests and forest-dwellers. Most of all, gratitude to forest-dwellers everywhere, who for millennia have protected the forests and our human heritage of knowing how to dwell in them.

Derrick Jensen:
My thanks go to Margo Baldwin for publishing, Helen Whybrow for editing, and Rowan Jacobsen for copyediting this book. I'd also like to thank Collette Fugere. All made important and clarifying suggestions. I thank John Osborn for teaching me how to be an activist.

This book could not have been written without the courage, wisdom, insight, patience, and humor of the trees who are my closest and dearest neighbors, nor without the frogs who sing their stories to me each night. My gratitude also is to the other creatures of the forest here, the coho salmon and Pacific lampreys, the beetles and solitary bees, the winged ants and the phoebes.

I'd like also to thank those everywhere who are putting their hearts and minds and bodies between the forests and those who would destroy them. If we can hold on only a little bit longer, civilization will collapse, and the forests and forest-dwellers can once again begin to live in peace, can begin to teach us all the lesson they've been teaching from the beginning: how to be human.

RESOURCES FOR GETTING INVOLVED

A SEED Europe
www.aseed.net
Action for Solidarity, Equality, the Environment and Development has campaigns on forests, gene technologies, free trade, multilateral development banks, energy, and human rights.

Amazon Watch
www.amazonwatch.org
Exposes the social and environmental impacts of industry in the Amazon Basin. Alerts public and private investors to the risks associated with controversial projects. Mobilizes technical, financial, legal, and public relations support for indigenous organizations fighting destructive projects.

American Lands Alliance
www.americanlands.org
Campaigns on fire, invasive species, off-road vehicles, old growth, roadless areas, forest restoration, Chile, and trade and forests.

Certified Wood Products Council
www.certifiedwood.org
Sources of certified wood and pulp. For recycled paper, see **ReThink Paper.**

Cultural Survival
www.cs.org
Promotes the rights, voices, and visions of indigenous peoples; publicizes land-rights campaigns in Colombia, Papua New Guinea, Taiwan, Malaysia, Nicaragua, Suriname, and elsewhere.

Dogwood Alliance
www.dogwoodalliance.org
Protects southern U.S. forests and communities by ending unsustainable industrial forest practices; stopping the construction of new or expanding wood-chipping facilities; passing forest protection policies in southern states; reducing the demand for virgin wood fiber, and encouraging the use of recycled and tree-free alternatives.

Endgame Clearinghouse on Forests and Corporations
www.endgame.org
Timber, oil, and mining corporation profiles, industry analyses, statistics on trade and deforestation, and links to environmental and community organizations.

FERN (Forests and the European Union Resource Network)
www.fern.org
Promotes the conservation and sustainable use of forests and respect for the rights of forest peoples. Campaigns include climate change, forest certification, export credit agencies, WTO trade agreements, intergovernmental agendas, aid and development cooperation, and the rights of forest peoples.

Forest Conservation Portal
www.forests.org
Vast database of news, analyses, and maps of the world's forests, dedicated to ending deforestation, preserving old-growth forests, sustainably managing other forests, maintaining climatic systems, and commencing the age of ecological restoration.

Forest Ethics
www.forestethics.org
Uses market power to educate individual consumers, large corporate purchasers, and distributors; and to protect endangered

forests, reform forestry practices, restore environmentally sensitive forests, increase decision-making power for indigenous communities, diversify forest-dependent economies, and reduce consumption of virgin wood fibers.

Forests Monitor
www.forestsmonitor.org
Investigates the timber industry to empower forest-dependent peoples and to show the links between trade, investment, and deforestation.

Friends of the Earth
www.foei.org
Campaigns include opposing the construction of pipelines in Peru, saving Australian koalas from logging, and confronting Dutch companies involved in Indonesian timber.

Global Forest Watch
www.globalforestwatch.org
Promotes transparency and accountability by tracking corporations, government agencies, and individuals that are sponsoring development activities and whether they are following national and local management laws and regulations.

Global Witness
www.oneworld.org/globalwitness
Exposes the links between environmental exploitation and human-rights abuses, including timber and conflicts in Zimbabwe, the Congo, Liberia, and Sierra Leone; illegal logging in Cambodia and Cameroon; and Japanese consumption.

Greenpeace
www.greenpeace.org
Forest campaigns include exposing illegal logging in the

Amazon, blockading European imports of African wood, and protecting the coastal rainforests of British Columbia.

Human Rights Watch
www.hrw.org
Recent reports cover Philippines human rights and forest management in the 1990s, corporations and human rights, labor abuse on banana plantations in Ecuador, complicity of U.S. oil corporations in human-rights violations in Indonesia, and resource exploitation and war in the Congo.

Indigenous Environmental Network
www.ienearth.org
Grassroots indigenous peoples whose mission is to protect the sacredness of Mother Earth from contamination and exploitation by strengthening, maintaining, and respecting traditional teachings and natural laws. Campaigns address mining and oil, energy, toxics, global warming, food, biodiversity, and sacred lands.

International Union for Conservation of Nature
www.iucn.org
Association of governments, NGOs, and scientists that publishes the official Red Data lists of threatened species.

Native Forest Network
www.nativeforest.org
Grassroots global network of forest activists, indigenous peoples, conservation biologists, and NGOs.

Native Web
www.nativeweb.org
Resources for indigenous peoples around the world. Links indigenous organizations and campaigns and documents such as

the Declaration of Brazilian Shamans Regarding Protection of Traditional Knowledge.

Oilwatch International
www.oilwatch.org.ec
Opposes the activities of oil companies in tropical countries. Recent issues of their bulletin *Tegantai* are devoted to oil and the military, Shel Oil Company, oil in mangroves, pipelines, oil and biodiversity, climate change, and human rights.

Project Underground
www.moles.org
Supports the human rights of communities resisting mining and oil exploitation, including the Ogoni in Nigeria, the Sarayacu Kichwa in Ecuador, and indigenous peoples in the Chiquitano forest in Bolivia.

Rainforest Action Network
www.ran.org
Campaigns include supporting the U'wa peoples fighting oil development in Colombia, pushing wood and paper manufacturers and retailers to phase out unsustainable products, and exposing the involvement of financial institutions such as the World Bank and Citigroup in global deforestation.

ReThink Paper
www.rethinkpaper.org
Information on and sources of recycled and tree-free paper.

Survival International
www.survival-international.org
Supports tribal peoples' right to decide their own future and helps them protect their lives, lands, and human rights. Recent campaigns address the rights of Bushmen in Botswana and

tribal recognition and land rights for Brazilian Indians affected by ranching and mining.

Taiga Rescue Network
www.sll.fi/TRN
NGOs, indigenous peoples, and individuals working for the protection and sustainable use of boreal forests.

Third World Network
www.twnside.org.sg
Southern coalition of environmental and human rights groups, with campaigns that address global trade rules and the resulting financial and economic crises, as well as biotechnology, biodiversity, indigenous knowledge, and community rights.

Transparency International
www.transparency.org
Global coalition against corruption; has initiated a Forest Integrity Network with the World Conservation Union and Harvard University's Center for International Development.

World Rainforest Movement
www.wrm.org.uy
Analysis, monitoring, and publicity, including an extensive directory of NGOs and indigenous peoples' organizations.

World Resources Institute
www.wri.org
Corporate-funded and technocratically-oriented, but WRI has published some solid reports and mapped the forest situation in various countries, and their Global Forest Watch and Forest Frontiers Initiative programs are worth monitoring.

NOTES

Deforestation

1. Murray Morgan, *The Last Wilderness* (Seattle, Wash.: University of Washington Press, 1976).
2. Greg Breining, "South China Tiger as Good as Extinct," *San Francisco Chronicle,* 9 January 2003, A1.
3. Public Citizen, "Corporate Welfare Examples in 1999," www.citizen.org/congress/welfare/articles.cfm?ID=1053, site visited January 23, 2003.
4. U.S. Forest Service, *1998 Report of the Forest Service,* www.fs.fed.us/pl/pdb/98report/02_stats.html, site visited January 23, 2003.
5. Farley Mowat, *Sea of Slaughter* (Toronto: Seal Books, 1989).
6. William Wade Keye, "Managing Forests, Protecting Watersheds," *San Francisco Chronicle,* December 1, 2002, D5.
7. Clive Ponting, *A Green History of the World: The Environment and the Collapse of Great Civilizations* (N. Y.: Penguin Books, 1991).
8. John Perlin, *A Forest Journey: An International Guide to Sustainable Forestry Practices* (Durango, Colo.: Kivaki Press, 1994), 35.
9. Ibid., 40.
10. Ibid., 35–39.
11. Ibid., 135–36.
12. D. Bryant, D. Nielsen, and L. Tangley, *The Last Frontier Forests: Ecosystems and Economies on the Edge* (D.C.: World Resources Institute, 1997), www.igc.org/wri/ffi/lff-eng/lff-toc.htm, site visited January 23, 2003.

Forest Dwellers

1. Margaret Atwood is quoted in Greenpeace, "Sorting through the Systems: A Greenpeace Briefing on Interfor's Forest Certifications,"www.greenpeaceusa.org/media/factshets/interfor_sorting_systems.pdf, site visited July 21, 2003.

Forest Dwellers, continued

2. International Union for the Conservation of Nature, Species Survival Commission, *1996 IUCN Red List of Threatened Animals* (Gland, Switzerland: IUCN, 1996), 36.
3. Bryant et al., *The Last Frontier Forests,* 17.
4. Margot Higgins, "Extinction Debt Come Due Long after Deforestation," Environmental News Network, www.enn.com/news/enn-stories/1999/10/101299/livingdead-26.html, site visited January 23, 2003.
5. Rainforest Foundation-US, "Why? Save the Rest," www.rainforestfoundation.org/why/html, site visited January 23, 2003.
6. Richard Drinnon, *Facing West: The Metaphysics of Indian-Hating & Empire-Building* (Norman, Okla.: University of Oklahoma Press, 1997), xiii.
7. Sylvia De Rooy, "Before the Wilderness," *Wild Humboldt* 1, Spring/Summer 2002, 12.
8. Marcus Colchester and Larry Lohmann, *The Struggle for Land and the Fate of the Forests* (Penang, Malaysia: World Rainforest Movement, *The Ecologist,* and Zed Books, 1993), 19.
9. Survival International, "Bushmen Silenced and Barred from Ancestral Lands," www.survival-international.org/bushman news020225.htm, site visited January 23, 2003.
10. Jeffrey St. Clair, "Panda Porn: The Grotesque Nuptials of WWF and Weyerhaeuser," *Anderson Valley Advertiser,* December 11, 2002, 5.
11. World Rainforest Movement, "Defenders of the Forest," www.wrm.org.uy/peoples.index.html, site visited January 23, 2003.

Accountability

1. Clay Geerdes, review of *Break Their Haughty Power: Joe Murphy in the Heyday of the Wobblies,* by Eugene Nelson, *Anderson Valley Advertiser,* April 24, 1996, 8.
2. Derrick Jensen, *The Culture of Make Believe* (N. Y.: Context Books, 2002).
3. Peter Dauvergne, *Shadows in the Forest: Japan and the Politics of Timber in Southeast Asia* (Cambridge, Mass.: MIT Press, 1997).

4. Jail Hurwitz, www.jailhurwitz.com, site visited January 23, 2003.
5. J.A. Savage, "Roots of Discontent," *Alternet,* October 22, 2002, www.alternet.org/story.html?StoryID=14356
6. Doug Bevington, "Confronting California's Logzilla," *Earth Island Journal* 15, online version (Winter 2000–2001), www.earthisland.org/ejournal/new_articles.cfm?articleID=67&journalID=43, site visited January 23, 2003.
7. A. Clay Thompson, "The King of Stumps: Red Emmerson's Sierra Pacific Industries Is the Single-Greatest Threat to California's Forests—and Until Recently, Nobody Was Paying Attention," *San Francisco Bay Guardian,* online version, June 28, 2000, www.sfbg.com/News/34/39/39stump.html, site visited January 23, 2003.
8. Mike Romano, "Who Killed the Timber Task Force?" *Seattle Weekly,* online version, July 9-15, 1998, www.seattleweekly.com/features/9827/features-romano.html, site visited January 23, 2003.

Killing Forests

1. B. H. Heede, "Response of a Stream in Disequilibrium to Timber Harvest," *Environmental Management* 15 (1991): 251–55; R. H. Rice, "A Perspective on the Cumulative Effects of Logging on Streamflow and Sedimentation," *Proceedings of the Edgebrook Conference, Berkeley, CA, July 2–3, 1980,* Special Publication No. 3268 (Berkeley, Calif.: University of California, Division Of Agricultural Science), 36–46.
2. D. Fredriksen and D. Harr, "Soil Vegetation, and Watershed Management," In *Forest Soils of the Douglas-Fir Region,* ed. Paul E Heilman et al. (Pullman, Wash.: Washington State University, 1981), 231–60.
3. Reed Noss, "Biodiversity in the Blue Mountains," (paper presented at the 1992 Blue Mountains Biodiversity Conference at Whitman College, Walla Walla, Wash., May 26–29, 1992).
4. W. D. Newmark, "Legal and Biotic Boundaries of Western North American National Parts," *Biological Conservation* 33 (1985): 197–208.

Killing Forests, continued

5. Reed Noss, *The Ecological Effects of Roads or The Road to Destruction* (Missoula, Mont.: Wildlands Center for Preventing Roads), www.wildlandscpr.org/resourcelibrary/reports/ecoleffectsroads.html.
6. Elliot Norse, *Ancient Forests of the Pacific Northwest* (D.C.: Island Press, 1990), 172–190; P. W. Adams and H. A. Froelich, *Compaction of Forest Soils* (Corvallis, Oreg.: Pacific Northwest Extension Service, 1981); M. P. Amaranthus and D. E. Steinfeld, "Soil Compaction after Yarding of Small-Diameter Douglas-Fir with a Small Tractor in Southwest Oregon," *U.S. Forest Service Resource Paper PNW-504,* 1997.
7. J. D. Knoepp, and W. T. Swand, " Long-term Effects of Commercial Sawlog Harvest on Soil Cation Concentrations," *Forest Ecology and Management* 93 (1997): 1–7.
8. Norse, *Ancient Forests of the Pacific Northwest,* 172–190.
9. W. R. Meehan, "Influence of Riparian Canopy on Macroinvertebrate Composition and Food Habits of Juvenile Salmonids in Several Oregon Streams," *U.S. Forest Service Resource Paper PNW-496,* 1996.
10. W. Nehlson, et al. "Pacific Salmon at the Crossroads," *Fisheries* 16 (1991): 4–21.
11. Larry D. Harris, *The Fragmented Forest: Island Biography Theory and the Preservation of Biotic Diversity* (Chicago: University of Chicago Press, 1984).
12. D. S. Wilcove, "Nest Predation in the Forest Tracts and the Decline of Migratory Songbirds," *Ecology* 66 (1985): 1211–14.
13. W. S. Alverson et al., "Forests Too Deer: Edge Effects in Northern Wisconsin," *Conservation Biology* 2, 1988.
14. A. L. Taylor and E. D. Forsmann, "Recent Range Extensions of the Barred Owl in Western America," *Condor* 78 (1976): 560–61.
15. Peter Morrison, *Ancient Forests on the Olympic National Forest* (Seattle, Wash.: Wilderness Society, 1990).
16. Chris Maser, *Redesigned Forest* (San Pedro, Calif.: R & E Miles, 1998).
17. Bryant, et al., *The Last Frontier Forests.*

18. U.S. Forest Service, "Pesticide Use Report," *1998 Report of the Forest Service,* Table 26.
19. W. R. Gast, et al., *Blue Mountain Forest Health Report,* U.S. Forest Service, Malheur, Umatilla, and Wallowa-Whitman National Forests, 1991.
20. Chris Maser, 'Logging to Infinity," *Anderson Valley Advertiser,* April 12, 1989, www.iww.org/iu120/local/Maser1.html.
21. W. Patricia Marchak, *Logging the Globe* (Montreal: McGill-Queens; University Press, 1995), 277–278; Sue Branford and Oriel Glock, *The Last Frontier: Fighting Over Land in the Amazon* (London: Zed Books, 1995), 118–119; Ricardo Carrere and Larry Lohmann, *Pulping the South: Industrial Tree Plantations and the World Paper Economy* (London: Zed Books, 1996), 138, 162–165.
22. Hillary Maycll, "Study Links Logging with Severity of Forest Fires," *National Geographic News,* December 3, 2001; F. Seigert, et al., "Increased Damage From Fires in Logged Forests During Droughts Caused by El Niño," *Nature* 414 (November 22, 2001): 437–40; Down to Earth/International Campaign for Ecological Justice in Indonesia. "Foreign Debt Fuels Forest Fires in Indonesia," August 2, 1999, lists.essential.org/stop-imf/msg 00198.html.
23. Tim Dodd, "Indonesian Forests in Aid Trade-Off," *Australian Financial Review,* 29 January 2000.
24. Down to Earth, "Foreign Debt Fuels Forest Fires in Indonesia."
25. Yarrow Robertson, "Briefing Document on Road Network Through Leuser Ecosystem," www.duke.edu/~myml/leuserBRIEFING.PDF.

Pulping the World

1. U.S. Senator Henry M. Keller, *Congressional Record,* February 26, 1909, 3226.
2. D. Austin, et al, *Contributions to Climate Change* (D.C.: World Resources Institute, 1998).
3. U.S. Forest Service, *Air Emissions from Wood and Wood-Based Products*. www.fpl.fs.fed.us/VOC/litature.htm, site visited July 21, 2003.

Pulping the Forests, continued

4. Maureen Smith, *The U.S. Paper Industry and Sustainable Production: An Argument for Restructuring* (Cambridge, Mass.: MIT Press, 1997), 111–113.
5. Ibid., 124.
6. Ibid., 121–122.
7. Alex J. Sagady, "Alert on Dioxin Emissions from Pulp and Paper Industry," National Wildlife Federation, December 18, 1996, lists.essential.org/1996/dioxin-1/msg00772.html, site visited January 23, 2003.
8. National Wildlife Federation, "EPA Announces New 'Cluster Rule'—Fails to Protect Americans from Toxic Paper Mill Pollution," November 14, 1997, www.nwf.org/watersheds/cluster.html, site visited January 23, 2003.
9. U.S. Environmental Protection Agency, *Pulp and Paper Industrial Sector Notebook,* www.csa.com/routenet/epan/pulppasn.html, site visited January 23, 2003.
10. Garden State EnviroNet, *Waste Reduction Tips for the Office,* www.gsenet.org/library/20rcy/offcrcyl.php, site visited January 23, 2003.
11. Smith, *The U.S. Paper Industry and Sustainable Production,* 159.
12. Ibid., 160–164.
13. Carrere and Lohman, *Pulping the South,* 24.
14. Michael Jaffe, *Industry Surveys: Paper & Forest Products* (N. Y.: Standard & Poor's, October 15, 1998), 10.
15. Carrere and Lohman, *Pulping the South,* 25.

Bodyguard of Lies

1. International Union for the Conservation of Nature, Species Survival Commission, *1996 IUCN Red List of Threatened Animals,* 36.
2. Georgia Forestry Association, *Myths & Facts: Forestry's Economic Impact on Georgia,* gfagrow.org/myths.htm, site visited July 21, 2003.
3. Bryant, et al., *The Last Frontier Forests,* 1, 20, 39.

4. George Draffan, *A Profile of the Weyerhaeuser Corporation* (Seattle, WA: Public Information Network, June 1999), www.endgame.org/weyerprofile.html, site visited January 23, 2003.
5. Plum Creek, www.plumcreek.com/company/our_forests.cfm, site visited December 26, 2002.
6. International Paper, www.internationalpaper.com/our_world/environment/forest/html, site visited December 26, 2002.
7. World Conservation Monitoring Centre, *Global Futures Bulletin,* November 1, 1998, and January 1, 1999.
8. Miller Freeman, *Pulp & Paper 1998 North American Factbook* (San Francisco: Miller Freeman, 1998), 72, 76; Smith, *The U.S. Paper Industry and Sustainable Production,* 40, 43, 72.
9. Jaffe, *Industry Surveys: Paper & Forest Products,* 4; Freeman, *Pulp & Paper 1998 North American Factbook,* 71
10. John A. Keslick, *Myths and Facts,* *www*.chesco.com/~treeman/nfpra/zerocut/maf.html, site visited July 21, 2003.

A Rigged System

1. U.S. Bureau of Corporations, *The Lumber Industry, Part I* (D.C.: U.S. Government Printing Office, 1913–14), 29.
2. Dee Brown, *Bury My Heart at Wounded Knee: An Indian History of the American West* (N.Y.: Holt, Rinehart, and Winston, 1970), 273, 449.
3. The statement about no human bodies lining the river was made by John Webster, editor of *Spokesman-Review* (Spokane, Washington) June 3, 1996.
4. Derrick Jensen, George Draffan, and John Osborn, *Railroads and Clearcuts: Legacy of Congress's 1864 Northern Pacific Railroad Land Grant* (Spokane, Wash.: Inland Empire Public Lands Council, 1995).
5. U.S. Forest Service, *1997 Resources Planning Act Assessment, Final Statistics* (D.C.: U.S. Forest Service, July 2000), Tables 12–15.
6. George Wuerthner, "Save the Forests: Let Them Burn," *High Country News,* August 29, 1988.

A Rigged System, continued

7. Ibid.
8. Ibid.
9. Sprig, "Green Polyester's Effect on Wildlife," *Earth First! Journal* (September/October 2002): 9.
10. John Osborn, "American Fire Policy: Conflagration—Total Exclusion—Restoration," *Transitions* (August/September 1992).
11. Ibid.
12. Ed Dorsch, "Federal Fire Sham: Logging and Fire Risk," *Forest Voice* (Fall 2002): 12; Hummingbird, "Dissent Flares Up Against Bush," *Earth First! Journal* (September/October 2002): 6; Wuerthner, "Save the Forests: Let Them Burn"; Sprig, "Green Polyester's Effect on Wildlife."

Corruption

1. Stephen Engelberg, "Tall Timber and the EPA," *New York Times,* May 21, 1995.
2. Draffan, *A Profile of the Weyerhaeuser Corporation*.
3. Ibid
4. Engelberg, "Tall Timber and the EPA."
5. Tom Paine, "Bush Appointments: Personnel Is Policy," Tom Paine Common Sense, www.tompaine.com/feature.cfm/ID/5261, site visited January 23, 2003.
6. Ibid.
7. Clearinghouse on Environmental Advocacy and Research, "Mark Rey Nominated for Undersecretary for Natural Resources and the Environment, Department of Agriculture," www.clearproject.org/reports_rey.html, site visited July 21, 2003.
8. Quoted in Andy Ryan, "Chainsaw Politics," *Seattle Weekly,* December 11–17, 2002, citing *National Journal,* 1997.
9. Ibid., 21.
10. St. Clair, "Panda Porn."
11. Public Employees for Environmental Responsibility, *Unindicted Co-conspirator: Timber Theft and the U.S. Forest Service,* (D.C.: PEER, March 1996), www.peer.org/publications/wp_coconspiresummary.html, site visited July 21, 2003.

12. Government Accountability Project, "National Forest Oversight Program" (D.C.: GAP), www.whistleblower.org/article.php?did=49&scid=83, site visited July 21, 2003.
13. Public Employees for Environmental Responsibility, *Unindicted Co-conspirator*.
14. Public Employees for Environmental Responsibility, *Stealing the Tongass: Playing by Alaska Rules in the U.S. Forest Service* (Washington, D.C.: PEER, November 1996), www.peer.org/publications/wp_tongasssummary.html, site visited July 21, 2003.
15. Public Employees for Environmental Responsibility, "Environmentalists ask Clinton Administration for Christmas: Defend the National Forests Against Corporate Timber Theft," (Washington, D.C.: PEER December 23, 1997), www.peer.org/action/pr_122397_timbertheft.html, site visited January 23, 2003.
16. American Forest & Paper Association, www.afandpa.org.
17. Common Cause, *Carrying a Big Stick: How Big Timber Triumphs in Washington* (Washington, D.C., Common Cause, 1997), www.commoncause.org/publications/timber.htm, site visited July 21, 2003.
18. Peter Dauvergne, *Shadows in the Forest*.
19. Friends of the Earth, "Briefing: European League Table of Imports of Illegal Tropical Timber," www.foe.co.uk/pubsinfo/briefings/html/2002090415432.html.
20. Colchester and Lohmann, *The Struggle for Land and the Fate of the Forests,* 104.
21. Bryant et al., *The Last Frontier Forests*, 20.
22. Colchester and Lohmann, *The Struggle for Land and the Fate of the Forests.*
23. Bryant et al., *The Last Frontier Forests*, 20.
24. Global Witness, *Made in Vietnam—Cut in Cambodia* (London: Global Witness, April 1999); Global Witness, *The Untouchables: Forest Crimes and the Concessionaires* (London: Global Witness, December 1999); Ker Munthit, "Cambodian Prime Minister Banishes UK-Based Environmental Group," *Associated Press,* December 24, 2002.

Globalization in the Real World

1. Abraham Lincoln, Letter to Colonel William F. Elkins, November 21, 1864, in *The Lincoln Encyclopedia,* by Archer H. Shaw (New York: Macmillan, 1950). Some scholars have questioned the authenticity of this quote; for example, see "Lincoln's Prophecy," Urban Legends References Pages, 66.165.133.65/quotes/lincoln.htm.
2. Branford and Glock, *The Last Frontier,* 127.
3. World Rainforest Movement and Forest Monitor, *High Stakes: The Need to Control Transnational Logging Companies: A Malaysian Case Study* (Ely, UK: Forests Monitor, August 1988), www.forestsmonitor.org/reports/highstakes/cover.htm, site visited July 21, 2003.
4. Ron Glastra, *Cut and Run: Illegal Logging and Timber Trade in the Tropics* (Ottawa: International Development Research Centre, 1999), 59–66.
5. U.S. General Accounting Office, *Promoting Democracy: Progress Report on U.S. Democratic Development Assistance to Russia,* Letter Report GAO/NSIAD-96-40, February 29, 1996.
6. Earth Liberation Prisioners, "Spirit of Freedom: Raul Zapatos," (London: Earth Liberation Prisioners), www.spiritoffreedom.org.uk/prisioners/raul.html, site visited July 21, 2003.
7. Glastra, *Cut and Run,* 67.
8. Teahjay J. Milton, "Logging Companies as Conduits for Domestic Political Repression," www.copla.org/teahjaypart1.htm, sited visited January 23, 2003.
9. *O'Dwyer's PR Daily,* December 31, 2002, www.odwyerpr.com/members/1231clsa_pipeline.htm, site visited January 23, 2003.
10. Glastra, *Cut and Run,* 48–49.
11. Former Secretary of Defense William Cohen, speech to Fortune 500 executives in Philadelphia, October 1998, cited in Daniel Schirmer, "President Clinton, A Corporate Offensive, and Okinawan Bases," (Cambridge, Mass.: Boston Okinawa Network, April 2000), www.boondocksnet.com/centennial/sctexts/schirmer2000.html.

12. Dauvergne, *Shadows in the Forest,* 1–2.
13. World Rainforest Movement, *Africa: Forests Under Threat* (Montevideo, Uruguay, August 2002).
14. Bryant, et al., *The Last Frontier Forests.*
15. Rainforest Relief, "Burma's Reign of Terror: Teak is Torture," (N.Y.: Rainforest Relief, March 24, 1997), www.forests.org/archive/asia/teakweek.htm, site visited July 21, 2003.

Consuming the World

1. Dennis Farney, "Unkindest Cut?" *Wall Street Journal,* June 18, 1990, A1.
2. Russell G. Coffee, *The Truth about Rainforest Destruction* (Austin, Tex.: Better Plant Press, 1996), 23.
3. PricewaterhouseCoopers, *2001 Global Forest and Paper Industry Survey* (New York: PricewaterhouseCoopers).
4. United Nations Food & Agriculture Organization, *Forest Products Yearbook 2000,* FAO Statistics Series 158 (Rome: UN FAO, 2002), A-6.
5. Joseph C. Tardiff, ed., *U.S. Industry Profiles: The Leading 100* (Detroit, Mich.: Gale Research, 2nd edition, 1998), 438, 443.
6. Wyng Chow, "U.S. Prime for Picking, Housing Suppliers Advised," *Vancouver Sun,* January 28, 2000.
7. W. Patricia Marchak, *Falldown: Forest Policy in British Columbia* (Vancouver, BC: David Suzuki Foundation and Ecotrust Canada, 1999), 86.
8. Clayoquot Rainforest Coalition, *The Costal Lumber Operations of British Columbia, Project 2: B.C. Mills to U.S. Markets*, Report II: 6–8 (Unpublished draft, August 1997).
9. Ibid., II:6, II:14.
10. Tom Green and Lisa Matthaus, *Cutting Subsides, or Subsidized Cutting?* Report commissioned by B.C. Coalition for Sustainable Forestry Solutions, July 12, 2001, iii–iv.
11. Glastra, *Cut and Run,* 4, 6.
12. Lynne Faltraco, "Chip Mill Fact Sheet," (Concerned Citizens of Rutherford County, N.C., January 2000).

Consuming the World, continued

13. Smith, *The U.S. Paper Industry and Sustainable Production,* 51, 159.
14. Carrere, *Pulping the South,* 38–40.
15. World Rainforest Movement and Forests Monitor, *High Stakes*.
16. Perlin, *A Forest Journey.*
17. Gordon Robinson, *The Forest and the Trees: A Guide to Excellent Forestry* (D.C.: Island Press, 1998), 18.
18. Charles Twining, *Phil Weyerhaeuser: Lumberman* (Seattle, Wash.: University of Washington Press, 1985), 135.
19. Smith, *The U.S. Paper Industry and Sustainable Production,* 27.
20. Carrere, *Pulping the South,* 52.
21. Neal G. Goulet and Amy Althoff, "New Freedom Wood Plant to Close; 225 Expect Layoffs," *New York Daily Record,* December 17, 1998, A1.
22. Carrere, *Pulping the South,* 29.
23. Phil Crawford, "Global Pulp and Paper Industry in Transition," *TAPPI Journal,* (January 1999): 45–47.
24. Smith, *The U.S. Paper Industry and Sustainable Production,* 254.
25. Carrere, *Pulping the South,* 89–90.
26. Jaffe, *Industry Surveys: Paper & Forest Products,* 4; Miller Freeman, *Pulp & Paper 1998 North American Factbook,* 71.
27. Tardiff, *U.S. Industry Profiles: The Leading 100,* 441.
28. David Korten, *When Corporations Rule the World* (West Hartford, Conn.: Kumarian Press and Berrett-Koehler, 1995), 79ff.
29. Al Stamborski, "Smurfit-Stone Will Cut 3,600 Jobs, Shuts Plants," *St. Louis Post-Dispatch,* November 25, 1998, C1.
30. Andrew Shain, "IP Plant Shutdown Continues," *Sun-News,* January 25, 1996, D1.
31. George Monbiot, "Murder in the Amazon," www.monbiot.com/dsp_article.cfm?article_id=300, site visited January 23, 2003.
32. Edward Tobin, IP Buying Union Camp In $6.6 Billion Deal," *Reuters,* November 24, 1998.
33. Official Board Markets (Chicago: Magazines for Industry, date unknown).

34. Martin Bayliss, *Asia Pacific Papermaker,* April 1998, 5.
5. *Pulp & Paper International,* March 1999, S19.
36. *Irish Company News,* November 1997, 35.
37. Brian Feagans, "Not Out of the Woods Yet: An Immediate Rescue Package is Needed for a Forest Industry that Cuts 30% of Its Jobs in 1996 as Domestic Demand for Timber Products Rose 16%," *Business Mexico,* June 1997, 48.
38. World Paper, "Offshore Fibre Search," *World Paper,* June 1995, 3.
39. Steve Anderson, "Shrinking Timber Supplies in Idaho Force Companies to Scout Offshore," *Idaho Business Review,* May 22, 1995, A1.
40. George Draffan, *Global Timber Titan: Weyerhaeuser* (Seattle, WA: Public Information Network, 1999), www.endgame.org/global-weyer.html, site visited July 21, 2003.
41. Antonia Juhasz, American Lands Alliance testimony to the U.S. Trade Representative's Office for the public hearing of the World Trade Organization, Washington, D.C., May 20–21, 1999, citing WT/CTE/W/67, November 7, 1997.
42. Ibid.
43. Mike Freeman, "World Buyers Due at Mill Auction," *Bulletin-Bend* (Oreg.), September 25, 1996, B1.
44. U.S. Overseas Private Insurance Corporation, "OPIC Board Approves Projects," (D.C.: OPIC, December 15, 1998).
45. Philip Hurst, *Rainforest Politics: Ecological Destruction in Southeast Asia* (London: Zed Books, 1990), 132.
46. Miller Freeman, *Forest Industries 1991–92 North American Factbook,* 178.
47. Environmental Investigation Agency, *Corporate Power, Corruption and the Destruction of the World's Forests* (London: EIA, 1996), 24–25.
48. Conrad B. MacKerron, *Business in the Rain Forests: Corporations, Deforestation, and Sustainability,* ed. Douglas G. Cogan (D.C.: Investor Responsibility Center, 1993), 56.
49. Environmental Investigation Agency, *Corporate Power, Corruption and the Destruction of the World's Forests,* 27, 41.

Consuming the World, continued

50. John Hancock, company news release, 13 October 1999.
51. Environmental Investigation Agency, *Corporate Power, Corruption and the Destruction of the World's Forests,* 22, 26, 28-29, 30-31, 40; World Rainforest Movement and Forest Monitor, *High Stakes;* Laurie Goering, "Gadding about Guyana," *Seattle Times,* August 10, 1997, K2.
52. Hurst, *Rainforest Politics,* 120.

The Failure of Solutions

1. is quoted in Gore Vidal, *The Decline and Fall of the American Empire* (Berkeley, Calif.: Odonian Press, 1992).
2. Nigel Sizer, "Perverse Habits," *World Resources Institute Forest Notes,* June 2000, 2.
3. United Nations Food and Agriculture Organization, *State of the World's Forests 1999* (Rome: Food and Agriculture Organization of the United Nations, 1999), 48.
4. Ibid., 41.
5. Greenpeace, *Chains of Destruction Leading from the World's Remaining Ancient Forests to the Japanese Market* (Amsterdam: Greenpeace International, April 2002), 1, 16.
6. Rainforest Relief, "NYC Council Introduces Legislation to Bar City's Use of Tropical Rainforests Hardwoods," News Release, February 18, 1998, www.rainforestrelief.org/newsnotes/nyc.htm, site visited January 23, 2003.
7. Carrere and Lohmann, *Pulping the South,* 17, 39–40; United Nations Food and Agriculture Organization, *State of the World's Forests 1999,* 48.
8. Norman Myers, *The Primary Source: Tropical Forests and Our Future* (N.Y.: W.W. Norton, 1985), 102–103.
9. Jaffe, *Industry Surveys: Paper & Forest Products,* 10.
10. Myers, *The Primary Source,* 102–103.
11. Hurst, *Rainforest Politics,* 95.
12. Smith, *The U.S. Paper Industry and Sustainable Production*, 97.
13. Ibid., 70, 89–91, 95.

14. Ibid., 125.
15. Ibid., 124ff.
16. Sizer, "Perverse Habits," 2.
17. Greenhouse Gas Protocol Initiative, www.ghgprotocol.org, site visited July 21, 2003.
18. Robert Repetto and Duncan Austin, *Pure Profit: The Financial Implications of Environmental Performance* (D.C.: World Resources Institute, 2000), vii.
19. Robert McIntyre, "On Tax Cuts, Loopholes and Avoidance," *Multinational Monitor,* June 2001, multinationalmonitor.org/mm2001/01june/june01interview.html, site visited July 21, 2003.
20. John J. Byckowski, "Firm's Plans Still Not in Stone," *Cincinnati Enquirer,* September 5, 1994, D1; Mike Boyer and Cliff Peale, "International Paper to add 400 jobs by mid-'97," *Cincinnati Post,* February 14, 1996, B6; "International Paper to cut 215," *Cincinnati Enquirer,* February 13, 1997, B16.
21. Hurst, *Rainforest Politics,* 117.
22. United Nations Food and Agriculture Organization, *Global Forest Resources Assessment 2000,* Main Report (Roma: FAO, 2001), 51, 57.
23. Rainforest Foundation/WALHI, "Environmentalists Challenge 'Eco-Timber' Go-Ahead for Logging in Endangered Tiger Habitat," News release, July 11, 2001.
24. Simon Counsell and Kim Terje Loraas, *Trading in Credibility: The Myth and Reality of the Forest Stewardship Council* (London: Rainforest Foundation-UK, November 20, 2002).
25. United Nations Food and Agriculture Organization, *Global Forest Resources Assessment 2000,* 413–422.
26. United Nations Food and Agriculture Organization, *State of the World's Forests 1999,* 35.
27. Michael Pilarski, ed., *Restoration Forestry: An International Guide to Sustainable Forestry Practices* (Durango, Colo.: Kivaki Press, 1994).
28. These definitions of community forestry are from Alistair Sarre, *The NSW Good Wood Guide* (Lismore, New South Wales, Australia: Rainforest Information Centre), www.rainforestinfo.org.au/good_wood/comm_fy.htm, site visited July 21, 2003.

The Failure of Solutions, continued

29. United Nations Conference on Environment and Development, *Rio Declaration on Environment and Development,* gopher://gopher.undp.org/00/unconfs/UNCED/English/riodecl, site visited July 21, 2003.
30. United Nations Conference on Environment and Development, *Rio Statement of Forest Princples,* gopher://gopher.undp.org/00/unconfs/UNCED/English/forestp, site visited July 21, 2003.
31. *Convention on Biological Diversity,* gopher://gopher.undp.org/00/unconfs/UNCED/English/biodiv, site visited July 21, 2003.
32. *Convention on the Conservation of Migratory Species of Wild Animals,* sedac.ciesin.org/entri/texts/migratory.wild.animals.1979.html, site visited July 21, 2003.
33. *Convention on International Trade in Endangered Species of Wild Fauna and Flora* (CITES), sedac.ciesin.org/entri/texts/cites.trade.endangered.species.1973.html, site visited January 23, 2003.
34. *International Tropical Timber Agreement,* sedac.ciesin.org/entri/texts/tropical.timber.1983.html, site visited January 23, 2003.
35. *Ramsar Convention on Wetlands of International Importance,* sedac.ciesin.columbia.edu:9080/entri/texts/migratory.wild.animals.1979.html, site visited July 21, 2003; United Nations Educational, Scientific and Cultural Organization, *Convention Concerning the Protection of the World Cultural and Natural Heritage,* whc.unesco.org/world_he.htm, site visited July 21, 2003. See also Harvard University Center for the Environment's "Index of Selected International Environmental Policy Resources by Policy Instrument," environment.harvard.edu/guides/intenvpol/indexes/treaty_TOC.html, site visited July 21, 2003.
36. Sizer, "Perverse Habits," 2, citing Lynch.
37. Americas Watch, *Rural Violence in Brazil* (N.Y.: Human Rights Watch, 1991).
38. Tom Barry and Deb Preusch, *Central America Fact Book* (N.Y.: Grove Press, 1986), 135: citing Jacobo Schatan, *La Agroindustria y el Sistema Centroamericano* (Mexico: CEPAL, 1983), 46.
39. Charles C. Geisler, "Ownership: An Overview," *Rural Sociology,* 1993, 58: 532-546.

40. Robert C. Fellmeth, *Power and Land in California* (D.C.: Center for Study of Responsive Law, 1971).
41. World Rainforest Movement, *Statement Drafted by Participants of the World Rainforest Movement Meeting in Penang, Malaysia on 14-17 April, 1989,* www.wrm.org.uy/statements/wrm.html, site visited July 21, 2003.
42. World Rainforest Movement, *The Montevideo Declaration: A Call for Action to Defend Forests and People Against Large-Scale Tree Monocrops,* June 1998, www.wrm.org.uy/statements/mvd.html, site visited July 21, 2003.
43. World Rainforest Movement, *Declaration to the World Summit on Sustainable Development and Forests* (Johannesburg, South Africa: August 2002), www.wrm.org.uy/statements/WSSD.html, site visited July 21, 2003.

Rejecting Gilgamesh

1. Plato's *Critias* as cited in Ponting, *A Green History of the World,* 76–77.
2. Brown, *Bury My Heart at Wounded Knee,* 273, 449.

BIBLIOGRAPHY

Adams, P. W. and H. A. Froelich. *Compaction of Forest Soils*. Corvallis, Oreg.: Pacific Northwest Extension Service, 1981.

Alverson, W. S., D. Waller and S. Solheim. "Forests Too Deer: Edge Effects in Northern Wisconsin," *Conservation Biology* 2, 1988.

Amaranthus, M. P. and D. E. Steinfeld. "Soil Compaction after Yarding of Small-Diameter Douglas-Fir with a Small Tractor in Southwest Oregon." *U.S. Forest Service Resource Paper PNW-504,* 1997.

Americas Watch. *Rural Violence in Brazil*. New York: Human Rights Watch, 1991.

Anderson, Steve. "Shrinking Timber Supplies in Idaho Force Companies to Scout Offshore." *Idaho Business Review,* May 22, 1995, A1.

Austin, D. et al. *Contributions to Climate Change*. Washington, D.C.: World Resources Institute, 1998.

Austin. City of Austin Green Building Program. *Sustainable Building Sourcebook: Wood Treatment*. www.txinfinet.com/sourcebook/woodtreatment.html, site visited 7/25/03.

Barry, Tom and Deb Preusch. *Central America Fact Book*. New York: Grove Press, 1986.

Bayliss, Martin. *Asia Pacific Papermaker,* April 1998.

Bevington, Doug. "Confronting California's Logzilla." *Earth Island Journal* 15, online version, Winter 2000–2001. www.earthisland.org/eijournal/new_articles.cfm?articleID=67&journalID=43, site visited 7/25/03.

Boyer, Mike and Cliff Peale. "International Paper to add 400 jobs by mid-'97." *Cincinnati Post*, February 14, 1996, B6.

Branford, Sue and Oriel Glock. *The Last Frontier: Fighting Over Land in the Amazon*. London: Zed Books, 1985.

Breining, Greg. "South China Tiger as Good as Extinct." *San Francisco Chronicle,* January 9, 2003, A1.

Brown, Dee. *Bury My Heart at Wounded Knee: An Indian History of the American West*. New York: Holt, Rinehart, and Winston, 1970.

Bryant, D., D. Nielsen, and L. Tangley. "The Last Frontier Forests: Ecosystems and Economies on the Edge." Washington, D.C.: World Resources Institute, 1997. www.igc.org/wri/ffi/lff-eng/lff-toc.htm, site visited 7/25/03.

Byczkowski, John J. "Firm's Plans Still Not in Stone." *Cincinnati Enquirer,* September 5, 1994, D1.

Carrere, Ricardo and Larry Lohmann. *Pulping the South: Industrial Tree Plantations and the World Paper Economy*. London: Zed Books, 1996.

Chow, Wyng. "U.S. Prime for Picking, Housing Suppliers Advised." *Vancouver Sun,* January 28, 2000.

Cincinnati Enquirer. "International Paper to Cut 215." *Cincinnati Enquirer,* February 13, 1997, B16.

Clayoquot Rainforest Coalition. *The Coastal Lumber Operations of British Columbia, Project 2: B.C. Mills to U.S. Markets*. Unpublished draft, August 1997.

Clearinghouse on Environmental Advocacy and Research. "Mark Rey Nominated for Undersecretary for Natural Resources and the Environment, Department of Agriculture." www.clearproject.org/reports_rey.html, site visited 7/25/03.

Coffee, Russell G. *The Truth About Rainforest Destruction*. Austin, Tex.: Better Planet Press, 1996.

Colchester, Marcus and Larry Lohmann. *The Struggle for Land and the Fate of the Forests*. Penang, Malaysia: World Rainforest Movement, *The Ecologist*, and Zed Books, 1993.

Common Cause. *Carrying a Big Stick: How Big Timber Triumphs in Washington*. Washington, D.C.: Common Cause, 1997. www.commoncause.org/publications/timber.htm, site visited 7/21/03.

Convention on Biological Diversity. gopher://gopher.undp.org/00/unconfs/UNCED/English/biodiv, site visited 7/21/03.

Convention on International Trade in Endangered Species of Wild Fauna and Flora (CITES). sedac.ciesin.org/entri/texts/cites.trade.endangered.species.1973.html, site visited January 23, 2003.

Convention on the Conservation of Migratory Species of Wild Animals. sedac.ciesin.org/entri/texts/migratory.wild.animals.1979.html, site visited 7/21/03.

Counsell, Simon and Kim Terje Loraas. *Trading in Credibility: The Myth and Reality of the Forest Stewardship Council.* London: Rainforest Foundation-UK, November 20, 2002. www.rainforestfoundationuk.org/FSC/RFA4REPORTfull.pdf

Crawford, Phil. "Global Pulp & Paper Industry in Transition." *TAPPI Journal,* January 1999, 45–47.

Dauvergne, Peter. *Shadows in the Forest: Japan and the Politics of Timber in Southeast Asia*. Cambridge, Mass.: MIT Press, 1997.

De Rooy, Sylvia. "Before the Wilderness." *Wild Humboldt* 1, Spring/Summer 2002.

Dodd, Tim. "Indonesian Forests in Aid Trade-Off." *Australian Financial Review,* January 29, 2000.

Dorsch, Ed. "Federal Fire Sham: Logging and Fire Risk." *Forest Voice,* Fall 2002, 12. www.forestcouncil.org/voice/issues/02_fall/wildfire4.php, site visited 7/21/03.

Down to Earth/International Campaign for Ecological Justice in Indonesia [UK]. *Foreign Debt Fuels Forest Fires in Indonesia,* August 2, 1999. lists.essential.org/stop-imf/msg00198.html.

Draffan, George. *Global Timber Titan: Weyerhaeuser*. Seattle, Wash.: Public Information Network, 1999. www.endgame.org/global-weyer.html, site visited 7/25/03.

Draffan, George. *Remaining Frontier Forest*. Seattle, Wash.: Public Information Network, date unknown. www.endgame.org/gtt-deforestation.html, site visited 7/25/03.

Drinnon, Richard. *Facing West: The Metaphysics of Indian-Hating & Empire-Building*. Norman, Okla.: University of Oklahoma Press, 1997.

Earth Liberation Prisioners. "Spirit of Freedom: Raul Zapatos." London: Earth Liberation Prisioners. www.spiritoffreedom.org.uk/prisioners/raul.html, visited 7/21/03.

Engelberg, Stephen. "Tall Timber And the E.P.A." *New York Times,* May 21, 1995.

Environmental Investigation Agency. *Corporate Power, Corruption and the Destruction of the World's Forests*. London: EIA, 1996.

Faltraco, Lynne. Chip Mill Fact Sheet. N.C.: Concerned Citizens of Rutherford County, January 2000.

Farney, Dennis. "Unkindest Cut?" *Wall Street Journal,* June 18, 1990, A1.

Feagans, Brian. "Not out of the Woods Yet: An Immediate Rescue Package is Needed for a Forest Industry that Cut 30% of Its Jobs in 1996 as Domestic Demand for Timber Products Rose 16%." *Business Mexico,* June 1997, 48.

Fellmeth, Robert C. *Power and Land in California: The Ralph Nader Task Force Report on Land Use in the State of California*. Washington, D.C.: Center For Study Of Responsive Law, 1971.

Forest Stewardship Council. www.fscoax.org, site visited 7/25/03.

Fredriksen, D. and D. Harr. "Soil, Vegetation, and Watershed Management." In *Forest Soils of the Douglas-fir Region,* edited by Paul E. Heilman et al. Pullman, Wash.: Washington State University, 1981, 231–60.

Freeman, Mike. "World Buyers Due at Mill Auction." *Bulletin-Bend* (Oreg.), September 25, 1996, B1.

Friends of the Earth. "Briefing: European League Table of Imports of Illegal Tropical Timber. www.foe.co.uk/pubsinfo/briefings/html/2002090415432.html.

Garden State EnviroNet. Waste Reduction Tips for the Office. www.gsenet.org/library/20rcy/offcrcyl.php, site visited 7/25/03.

Gast, W.R. et al. *Blue Mountain Forest Health Report*. U.S. Forest Service, Malheur, Umatilla, and Wallowa-Whitman National Forests, 1991.

Geerdes, Clay. Book review of *Break Their Haughty Power: Joe Murphy in the Heyday of the Wobblies*. *Anderson Valley Advertiser,* April 24, 1996, 8.

Geisler, Charles C. "Ownership: An Overview." *Rural Sociology*, 1993, 58: 532-546.

George Draffan. *A Profile of the Weyerhaeuser Corporation*. Seattle, Wash.: Public Information Network, June 1999. www.endgame.org/weyerprofile.html, site visited 7/21/03.

Georgia Forestry Association. *Myths & Facts: Forestry's Economic Impact on Georgia*. gfagrow.org/myths.htm, site visited 7/21/03.

Glastra, Ron. *Cut and Run: Illegal Logging and Timber Trade in the Tropics*. Ottawa: International Development Research Centre, 1999.

Global Witness. *Made in Vietnam—Cut in Cambodia*. London: Global Witness, April 1999. www.globalwitness.org/reports/download.php/00061, site visited 7/21/03.

Global Witness. *The Untouchables: Forest Crimes and the Concessionaires*. London: Global Witness, December 1999. www.globalwitness.org/reports/index.php?section=cambodia

Goering, Laurie. "Gadding about Guyana." *Seattle Times,* August 10, 1997, K2. archives.seattletimes.nwsource.com/cgi-bin/texis.cgi/web/vortex/display?slug=guya&date=19970810, site visited 7/21/03.

Goulet, Neal G. and Amy Althoff. "New Freedom Wood Plant to Close; 225 Expect Layoffs." *New York Daily Record,* December 17, 1998, A1.

Government Accountability Project. "National Forest Oversight Program." Washington, D.C.: GAP, date unknown. www.whistleblower.org/article.php?did=49&scid=83, site visited 7/21/03.

Green, Tom and Lisa Matthaus. *Cutting Subsidies, or Subsidized Cutting?* Report Commissioned by B.C. Coalition for Sustainable Forestry Solutions, July 12, 2001.

Greenhouse Gas Protocol Initiative. www.ghgprotocol.org/, site visited 7/21/03.

Greenpeace. "Sorting through the Systems: A Greenpeace Briefing on Interfor's Forest Certifications." www.greenpeaceusa.org/media/factshets/interfor_sorting_systems.pdf, site visited 7/25/03.

Greenpeace. *Chains of Destruction Leading from the World's Remaining Ancient Forests to the Japanese Market*. Amsterdam: Greenpeace International, April 2002.

Harris, Larry D. *The Fragmented Forest: Island Biogeography Theory and the Preservation of Biotic Diversity*. Chicago: University of Chicago Press, 1984.

Heede, B.H. "Response of a Stream in Disequilibrium to Timber Harvest." *Environmental Management* 15, 1991, 251–55.

Higgins, Margot. "Extinction Debts Come Due Long after Deforestation." *ENN: Environmental News Network*, October 12, 1999. www.enn.com/news/enn-stories/1999/10/101299/livingdead_6346.asp, site visited 7/25/03.

Hummingbird. "Dissent Flares Up Against Bush." *Earth First! Journal*, September/October 2002, 6.

Hurst, Philip. *Rainforest Politics: Ecological Destruction in Southeast Asia*. London: Zed Books, 1990.

International Paper www.internationalpaper.com/our_world/environment/forest.html , site visited December 26, 2002.

International Tropical Timber Agreement. sedac.ciesin.org/entri/texts/tropical.timber.1983.html, site visited January 23, 2003.

International Union for the Conservation of Nature, Species Survival Commission. *1996 IUCN Red List of Threatened Animals*. Gland, Switzerland: IUCN, 1996.

Irish Company News, November 1997, 35.

Jail Hurwitz. www.jailhurwitz.com, site visited 7/25/03.

Jensen, Derrick, George Draffan, and John Osborn. *Railroads and Clearcuts: Legacy of Congress's 1864 Northern Pacific Railroad Land Grant*. Spokane, Wash.: Inland Empire Public Lands Council, 1995.

Jensen, Derrick. *The Culture of Make Believe*. New York: Context Books, 2002.

Johannesburg. www.wrm.org.uy/statements/WSSD.html, site visited 7/25/03.

John Hancock. Company news release. October 13, 1999.

Johnston, David Robert. *Building Green in a Black and White World, Chapter* 2. Housing Zone. www.housingzone.com/topics/nahb/green/nhb00ca022.asp, site visited 7/25/03.

Juhasz, Antonia. American Lands Alliance testimony to the U.S. Trade Representative's Office for the public hearing on the World Trade Organization, Washington, D.C., May 20–21, 1999, citing WT/CTE/W/67, November 7, 1997.

Keller, U.S. Senator Henry M. *Congressional Record,* February 26, 1909, 3226

Keslick, John A. *Myths and Facts*. www.chesco.com/~treeman/nfpra/zerocut/maf.html, site visited 7/21/03.

Keye, William Wade. "Managing Forests, Protecting Watersheds." *San Francisco Chronicle,* December 1, 2002, D5.

Knoepp, J. D. and W. T. Swank. "Long-term Effects of Commercial Sawlog Harvest on Soil Cation Concentrations." *Forest Ecology and Management* 93, 1997, 1–7.

Korten, David. *When Corporations Rule the World*. West Hartford, Conn.: Kumarian Press and Berrett-Koehler, 1995 .

Lincoln, Abraham. Letter to Colonel William F. Elkins, November 21, 1864. In *The Lincoln Encyclopedia,* by Archer H. Shaw. New York: Macmillan, 1950.

Lynch, O. *Securing Community-Based Tenurial Rights in the Tropical Forests of Asia*. Washington, D.C.: World Resources Institute, 1992.

MacKerron, Conrad. *Business in the Rainforests: Corporations, Deforestation and Sustainability*. Washington D.C.: Investor Responsibility Research Center, 1993, 56.

Marchak, W. Patricia. *Falldown: Forest Policy in British Columbia*. Vancouver, B.C.: David Suzuki Foundation and Ecotrust Canada, 1999.

Marchak, W. Patricia. *Logging the Globe*. Montreal: McGill-Queens' University Press, 1995.

Maser, Chris. "Logging to Infinity." *Anderson Valley Advertiser,* April 12, 1989. www.iww.org/iu120/local/Maser1.html, site visited 7/25/03.

Maser, Chris. *The Redesigned Forest*. San Pedro, Calif.: R & E Miles, 1988.

Mayell, Hillary. "Study Links Logging with Severity of Forest Fires." *National Geographic News,* December 3, 2001.

McIntyre, Robert. "On Tax Cuts, Loopholes and Avoidance." *Multinational Monitor,* June 2001. multinationalmonitor.org/mm 2001/01june/june01interview.html , site visited 7/21/03.

Meehan, W.R. "Influence of Riparian Canopy on Macroinvertebrate Composition and Food Habits of Juvenile Salmonids in Several Oregon Streams." *U.S. Forest Service Resource Paper* PNW-496, 1996.

Michael Jaffe, *Industry Surveys: Paper & Forest Products*. New York: Standard & Poor's, October 15, 1998.

Miller Freeman. *Forest Industries 1991–92 North American Factbook*. San Francisco: Miller Freeman, 178.
Miller Freeman. *Pulp & Paper 1998 North American Factbook*. San Francisco: Miller Freeman, 1998.
Milton, Teahjay, J. "Logging Companies as Conduits for Domestic Political Repression." www.copla.org/teahjaypart1.htm, site visited 7/25/03.
Monbiot, George. "Murder in the Amazon." www.monbiot.com/dsp_article.cfm?article_id=300, site visited 7/25/03.
Morgan, Murray. *The Last Wilderness*. Seattle, Wash.: University of Washington Press, 1976.
Morrison, Peter. *Ancient Forests on the Olympic National Forest*. Seattle, Wash.: Wilderness Society, 1990.
Mowat, Farley. *Sea of Slaughter*. Toronto: Seal Books, 1989.
Munthit, Ker. "Cambodian Prime Minister Banishes UK-Based Environmental Group." Associated Press, December 24, 2002.
Myers, Norman. *The Primary Source: Tropical Forests and Our Future*. New York: W.W. Norton, 1985.
National Wildlife Federation. "EPA Announces New 'Cluster Rule'—Fails to Protect Americans from Toxic Paper Mill Pollution," November 14, 1997. www.nwf.org/watersheds/cluster.html, site visited 7/21/03.
Nehlson, W. et al. "Pacific Salmon at the Crossroads." *Fisheries* 16, 1991, 4–21.
Newmark, W. D. "Legal and Biotic Boundaries of Western North American National Parks." *Biological Conservation* 33, 1985, 197–208.
Niemi, Arnie et al. *Salmon, Timber, and the Economy*. Eugene, Oreg.: ECONorthwest, December 1999.
Norse, Elliot. *Ancient Forests of the Pacific Northwest*. Washington, D.C.: Island Press, 1990.
North Coast Journal. "Tribe Upset over Klamath." *North Coast Weekly Journal,* October 10, 2002. www.northcoastjournal.com/101002/news1010.html, site visited 7/25/03.
Noss, Reed. "Biodiversity in the Blue Mountains." Paper presented at the 1992 Blue Mountains Biodiversity Conference, Whitman College, Walla Walla, Wash., May 26–29, 1992.
Noss, Reed. *The Ecological Effects of Roads or The Road To Destruction*. Missoula, Mont.: Wildlands Center for Preventing Roads, date unknown. www.wildlandscpr.org/resourcelibrary/reports/ecoleffectsroads.html.

O'Dwyer's PR Daily, December 31, 2002. www.odwyerpr.com/members/1231clsa_pipeline.htm, site visited 7/25/03.

Official Board Markets. Chicago: Magazines for Industry, date unknown.

Osborn, John. "American Fire Policy: Conflagration—Total Exclusion—Restoration." *Transitions*, August/September 1992.

Pan European Forest Certification Council. www.pefc.org, site visited 7/25/03.

Perlin, John. *A Forest Journey: The Role of Wood in the Development of Civilization*. Cambridge, Mass.: Harvard University Press, 1989.

Pilarski, Michael, ed. *Restoration Forestry: An International Guide to Sustainable Forestry Practices*. Durango, Colo.: Kivaki Press, 1994.

Plum Creek. www.plumcreek.com/company/our_forests.cfm, site visited December 26, 2002.

Ponting, Clive. *A Green History of the World: The Environment and the Collapse of Great Civilizations*. New York: Penguin Books, 1991.

PricewaterhouseCoopers. *2001 Global Forest & Paper Industry Survey*. New York: PricewaterhouseCoopers.

Public Citizen. "Corporate Welfare Examples for 1999." Washington, D.C.: Citizen. www.citizen.org/congress/welfare/articles.cfm?ID=1053, site visited 7/25/03.

Public Employees for Environmental Responsibility. "Environmentalists ask Clinton Administration for Christmas: Defend the National Forests Against Corporate Timber Theft." Washington, D.C.: PEER, December 23, 1997. www.peer.org/action/pr_122397_timbertheft.html, site visited January 23, 2003.

Public Employees for Environmental Responsibility. *Stealing the Tongass: Playing by Alaska Rules in the U.S. Forest Service*. Washington, D.C.: PEER, November 1996. www.peer.org/publications/wp_tongasssummary.html, site visited 7/21/03.

Public Employees for Environmental Responsibility. *Unindicted Co-conspirator: Timber Theft and the U.S. Forest Service.* Washington, D.C.: PEER, March 1996. www.peer.org/publications/wp_coconspiresummary.html, site visited 7/21/03.

Pulp & Paper International, March 1999, S19.

Rainforest Foundation-US. "Why? Save The Rest." www.rainforestfoundation.org/why.html, site visited 7/25/03.

Rainforest Foundation/WALHI. "Environmentalists Challenge 'Eco-Timber' Go-Ahead for Logging in Endangered Tiger Habitat." News release, July 11, 2001.

Rainforest Relief. "Burma's Reign of Terror: Teak is Torture." New York: Rainforest Relief, March 24, 1997. www.forests.org/archive/asia/teakweek.htm, site visited 7/21/03.

Rainforest Relief. "NYC Council Introduces Legislation to Bar City's Use of Tropical Rainforest Hardwoods." News release, February 18, 1998. www.rainforestrelief.org/newsnotes/nyc.htm, site visited 7/25/03.

Ramsar Convention on Wetlands of International Importance. sedac.ciesin.columbia.edu:9080/entri/texts/migratory.wild.animals.1979.html

Raphael, Ray. *Tree Talk: The People and Politics of Timber*. Washington, D.C.: Island Press, 1981.

Repetto, Robert and Duncan Austin. *Pure Profit: The Financial Implications of Environmental Performance*. Washington, D.C.: World Resources Institute, 2000, vii .

Rice, R.H. "A Perspective on the Cumulative Effects of Logging on Streamflow and Sedimentation." In *Proceedings of the Edgebrook Conference, Berkeley, CA, July 2–3, 1980.* Berkeley: University of California, Division of Agricultural Science, Special Publication No. 3268, 36–46.

Robertson, Yarrow. Briefing Document on Road Network Through Leuser Ecosystem. www.duke.edu/~myml/LeuserBRIEFING.PDF.

Robinson, Gordon. *The Forest and the Trees: A Guide to Excellent Forestry*. Washington, D.C.: Island Press, 1988.

Romano, Mike. "Who Killed the Timber Task Force?" *Seattle Weekly,* online version, July 9–15, 1998. www.seattleweekly.com/features/9827/features-romano.shtml, site visited 7/25/03.

Ryan, Andy. "Chainsaw Politics." *Seattle Weekly,* December 11–17, 2002. www.seattleweekly.com/features/0250/news-ryan.shtml

Sagady, Alex J. "Alert on Dioxin Emissions from Pulp and Paper Industry." National Wildlife Federation, December 18, 1996. lists.essential.org/1996/dioxin-l/msg00772.html, site visited 7/25/03.

Sarre, Alistair. *The NSW Good Wood Guide*. Lismore, New South Wales, Australia: Rainforest Information Centre, www.rainforestinfo.org.au/good_wood/comm_fy.htm, site visited 7/21/03.

Savage, J.A. "Roots of Discontent." Alternet, October 22, 2002 www.alternet.org/story.html?StoryID=14356

Schatan, Jacobo. *La Agroindustria y el Sistema Centroamericano*. Mexico: CEPAL, 1983.

Schirmer, Daniel. "President Clinton, A Corporate Offensive, and Okinawan Bases." Cambridge, Mass.: Boston Okinawa Network,

April 2000. www.boondocksnet.com/centennial/sctexts/schirmer 2000.html site visited 7/21/03.

Shain, Andrew. "IP Plant Shutdown Continues." *Sun-News* (Myrtle Beach, S.C.), January 25, 1996, D1.

Siegert, F., G. Ruecker, A. Hinrichs and A. A. Hoffmann. "Increased damage from fires in logged forests during droughts caused by El Niño." *Nature* 414, November 22, 2001, 437–440.

Sizer, Nigel. "Perverse Habits." *World Resources Institute Forest Notes,* June 2000, 2.

Smith, Maureen. *The U.S. Paper Industry and Sustainable Production: An Argument for Restructuring*. Cambridge, Mass.: MIT Press, 1997.

Sprig. "Green Polyester's Effect on Wildlife." *Earth First! Journal,* September/October 2002, 9.

St. Clair, Jeffrey. "Panda Porn: The Grotesque Nuptials of WWF and Weyerhaeuser." *Anderson Valley Advertiser,* December 11, 2002, 5.

Stamborski, Al. "Smurfit-Stone Will Cut 3,600 Jobs, Shut Plants." *St. Louis Post-Dispatch,* November 25, 1998, C1.

Survival International. "Bushmen Silenced and Barred from Ancestral Lands." www.survival-international.org/bushmannews020225.htm, site visited 7/25/03.

Tardiff, Joseph C., editor. *U.S. Industry Profiles: The Leading 100*. Detroit: Gale Research, 2nd edition, 1998.

Taylor, A.L. and E.D. Forsmann. "Recent Range Extensions of the Barred Owl in Western America." *Condor* 78, 1976, 560–61.

Thompson, A. Clay. "The King of Stumps: Red Emmerson's Sierra Pacific Industries Is the Single-Greatest Threat to California's Forests—and Until Recently, Nobody Was Paying Attention." *San Francisco Bay Guardian,* online version, June 28, 2000. www.sfbg.com/News/34/39/39stump.html, site visited 7/25/03.

Tobin, Edward. "IP Buying Union Camp In $6.6 Billion Deal." *Reuters,* November 24, 1998.

Tom Paine. "Bush Appointments: Personnel Is Policy." Tom Paine.common sense. www.tompaine.com/feature.cfm/ID/5261, site visited 7/25/03.

Twining, Charles. *Phil Weyerhaeuser: Lumberman*. Seattle: University of Washington Press, 1985.

U.S. Bureau of Corporations. *The Lumber Industry*. Washington, D.C.: U.S. Government Printing Office, 1913-14.

U.S. Environmental Protection Agency. *Pulp and Paper Industrial Sector Notebook*. www.csa.com/routenet/epan/pulppasn.html, site visited 7/25/03.

U.S. Forest Service, "Pesticide Use Report." In *1998 Report of the Forest Service*, Table 26. www.fs.fed.us/pl/pdb/98report/table_26.html, site visited 7/21/03.

U.S. Forest Service. *1997 Resources Planning Act Assessment, Final Statistics*. Washington, D.C.: U.S. Forest Service, July 2000.

U.S. Forest Service. *1998 Report of the Forest Service*. www.fs.fed.us/pl/pdb/98report/02_stats.html, site visited 7/25/03.

U.S. Forest Service. *Air Emissions from Wood and Wood-Based Products*. www.fpl.fs.fed.us/VOC/litature.htm, site visited 7/21/03.

U.S. General Accounting Office. *Promoting Democracy: Progress Report on U.S. Democratic Development Assistance to Russia*. Letter Report GAO/NSIAD-96-40, February 29, 1996. Published at www.fas.org/man/gao/ns96040.htm, site visited 7/21/03.

U.S. Overseas Private Insurance Corporation. "OPIC Board Approves Projects." Washington, D.C.: OPIC, December 15, 1998.

United Nations Conference on Environment and Development. *Rio Statement of Forest Princples*. gopher://gopher.undp.org/00/unconfs/UNCED/English/forestp, site visited 7/21/03.

United Nations Conference on Environment and Development. *Rio Declaration on Environment and Development*. gopher://gopher.undp.org/00/unconfs/UNCED/English/riodecl, site visited 7/21/03.

United Nations Educational, Scientific and Cultural Organization. *International Conference on Biosphere Reserves (Seville)*. www.unesco.org/mab/publications/brbullet/br3_02a.htm, sited visited 7/25/03.

United Nations Educational, Scientific and Cultural Organization. *Convention Concerning the Protection of the World Cultural and Natural Heritage*. whc.unesco.org/world_he.htm

United Nations Food and Agriculture Organization. *Global Forest Resources Assessment 2000*. UN FAO Forestry Paper 140, 51. Rome: United Nations Food and Agriculture Organization. www.fao.org/forestry/fo/fra/docs/main.

United Nations Food and Agriculture Organization. *State of the World's Forests 1999*. Rome: UN FAO.

United Nations Food and Agriculture Organization. *Yearbook of Forest Products 2000*. FAO Statistics Series 158. Rome: United Nations Food and Agriculture Organization, 2002.

Vidal, Gore. *The Decline and Fall of the American Empire*. Berkeley, Calif.: Odonian Press, 1992.

Weyerhaeuser. 1998 US Securities & Exchange Commission Form 10-K.

Wilcove, D.S.. "Nest Predation in Forest Tracts and the Decline of Migratory Songbirds." *Ecology* 66, 1985, 1211–14.

World Bank Group. *Revised Forest Strategy for the World Bank Group*. Draft Report, July 30, 2001.

World Conservation Monitoring Centre. *Global Futures Bulletin,* November 1, 1998, and January 1, 1999.

World Paper. "Offshore Fibre Search." *World Paper*, June 1995, 3.

World Rainforest Movement and Forests Monitor. *High Stakes: The Need to Control Transnational Logging Companies: A Malaysian Case Study*. Ely, UK: Forests Monitor, August 1988. www.forestsmonitor.org/reports/highstakes/cover.htm site visited 7/21/03.

World Rainforest Movement. "Defenders of the Forest." www.wrm.org.uy/peoples/index.html site visited 7/25/03.

World Rainforest Movement. *Africa: Forests Under Threat*. Montevideo, Uruguay, August 2002.

World Rainforest Movement. *Declaration to the World Summit on Sustainable Development and Forests*. Johannesbug, South Africa, August 2002. www.wrm.org.uy/statements/WSSD.html, site visited 7/21/03.

World Rainforest Movement. *Statement Drafted by Participants of the World Rainforest Movement Meeting in Penang, Malaysia on 14-17 April, 1989*. www.wrm.org.uy/statements/wrm.html, site visited 7/25/03.

World Rainforest Movement. *The Montevideo Declaration: A Call for Action to Defend Forests and People Against Large-Scale Tree Monocrops, June 1998*. www.wrm.org.uy/statements/mvd.html, site visited 7/21/93

Wuerthner, George. "Save the Forests: Let Them Burn." *High Country News,* August 29, 1988.

INDEX

ABOUT THE AUTHORS

George Draffan has worked as a carpenter and corporate librarian, and been a volunteer forest activist and freelance researcher and writer. He is the author or co-author of *Cascadia Wild, Railroads and Clearcuts, A Primer on Corporate Power,* and *The Elite Consensus*. Some of his work can be seen at the Public Information Network website at www.endgame.org.

Derrick Jensen is the author of *A Language Older than Words, Listening to the Land, The Culture of Make Believe,* and *Railroads and Clearcuts.* He lives in northern California, where he works to protect and to rehabilitate forest and stream habitat. Some of his work can be seen at www.derrickjensen.org.

Vandana Shiva is the author of *Stolen Harvest, The Violence of the Green Revolution, Water Wars,* and many other books and articles. She one of India's leading physicists, an internationally renowned activist, and winner of the Right Livelihood Award, known as "the alternative Nobel Prize." She is the founder and Director of Research Foundation for Science Technology and Natural Resource Policy.

As well as being a publisher of books for sustainable living, Chelsea Green is committed to being a sustainable business enterprise. This means reducing natural resource and energy use to the maximum extent possible. We print our books and catalogs on chlorine-free recycled paper, using soy-based inks, whenever possible. *Strangely Like War* was printed on Legacy Trade Book Natural, a 100% post-consumer waste recycled paper supplied by Webcom.